# DISGUISED AS THE DEVIL

# DISGUISED AS THE DEVIL:
## How Lyme Disease Created Witches and Changed History

M. M. Drymon

Wythe Avenue Press
Brooklyn
New York

Copyright © 2008 by M.M. Drymon
All rights reserved.
Published by Wythe Avenue Press.
www.wythepress.com
Design S.L. Powers
Graphics N.W. D. III Studio


Library of Congress Cataloging Data
     Drymon, M. M.
Disguised as the devil: how Lyme disease created witches and changed history/ by M.M. Drymon
p. cm.
Includes bibliographical references and index.
ISBN 978-0-6152-0061-3
1.Lyme disease  2. Witchcraft- History- Colonial Period 1600-1775
I. Title
2008926184


Manufactured in the United States of America
First Edition

*For*
Goodwife Rebecca Chamberlain
*of Billerica, Massachusetts,*
*who died in prison at Cambridge on September 26, 1692,*
*accused of practicing witchcraft.*


A portion of the proceeds from this work will go towards Lyme disease research and advocacy.

# CONTENTS

INTRODUCTION   1
I. AFFLICTION   31
II. PREDICTING RISK   77
A COMPARISON   97
THINKING IN TIME   104
III. ENGLISH SETTLEMENT IN THE NEW WORLD   105
IV. WAS IT LYME DISEASE ?   117
V. MASSACHUSETTS BAY: SCURVY DÉJÀ VU?   125
VI. THE PIG IN THE PROMISED LAND   137
WHY WOMEN?   150
VII. THE WITCH AT THE EDGE OF THE WOODS   151
VIII. SALEM VILLAGE   175
IX. PLACE, POLITICS AND SOCIAL CONTEXT   187
X. CONCLUSIONS   203

# APPENDIXES

A. Seventeenth Century Clothing   221
B. English Cultural Practices   223
C. Vitamin C Content of Foods/ Antimicrobial Activity of Traditional Spices   229
D. Moisture and Incidence: The Witch Drought   231
E. European Tick Infection rates   235
F. Grounds for the Examination of a Witch   237
G. Timeline/The Afflicted/Geographic Area   239
H. Population Change Model   259
J. Lyme disease diagnostic criteria   260
SELECTED BIBLIOGRAPY   263
INDEX   275
ACKNOWLEDGMENTS   285

## ILLUSTRATIONS

Figures

1. *Dirty Truth About Lyme Disease,* 4
   Published May 20, 1999, by permission Fairfield County Weekly, Connecticut
2. Mr. Preventionguy Poster, www.lymeassociation.org 9
3. Chart A: Seasonality of Ticks 23
4. Enlarged view of *Ixodes* Ticks on Fingertip 24
5. Factors Associated with Lyme disease Risk 25
6. Chart B: Reported Cases, 1994-2002, CDC, and
   Chart C: Witch Trial Participants 26
7. Present/Past Characteristics 30
8. Life Cycle of *Ixodes scapularis* Ticks, CDC 33
9. Simplified Population Bottleneck Diagram 34
10. Wooded Hillside in Modern Danvers 36
11. *Borrelia burgdorferi* in Spirochete Form 44
12. *Borrelia burgdorferi* Converted to Cyst Form 45
13. *Peromyscus leucopus*: the White Footed Mouse 47
14. *Odocoileus virginianus*: the White Tailed Deer 47
15. Rash on Ankle and Woodcut of Devil's Mark on Ankle 59
16. Examples of *Erythema Migrans* Rash 60
17. Attached tick 61
18. 17th Century Pins 62
19. Enlarged View of Ticks Next to Modern Pin 63
20. A Swollen Gland Can Look Like an Egg 64
21. Bell's Palsy 65
22. Illustration of a "*fit*," J.M. Charcot 68
23. Soil Texture Study, Wisconsin 78
24. Land Cover Categories 83
25. Landscape Characteristics: High Risk or Low Risk 85
26. Drag Collecting 'Flag' 119
27. Photograph of Scurvy *Macula* 120
28. Photograph of ECM Rash on an Ankle 120
29. Total Biomass Consumed in 17th C. Boston 134
30. A Naked Witch Riding on a Boar from *Melancholy* by 137
    Lucas Cranach, the Elder, 1530
31. A Pig, from *The Swimming of Mary Sutton*, 1613 141
32. 14th C. Depiction of Pigs During Pannage 142
33. Woodcut Portrait of a Witch, 1643 151
34. Weather Witches, *De Lamiis et Phitonicis Mulieribus*, 1493 154
35. Frontispiece of *Malleus Maleficarum.* 156

*Figures, cont'd.*
36. Witch Kissing the Anus of the Devil from      157
    *The Compendium Maleficarum*
37. *Hunters in the Snow* by Pieter Bruegel, the Elder,      159
    completed in February 1565
38. Woodcut of Witches with their Familiars from      162
    *The Wonderful Discoverie of the Witchcrafts of*
    *Margaret and Phillip Flower*, 1619
39. Broadside Advertisement of Witches Being Burned,      163
    Durneburg, Germany, 1555
40. A 17th C. Glass Bottle      166
41. *Witches Flying up the Chimney* by Gillotole de Givry, 1579      167
42. Frontispiece of a Pamphlet, 1603      170
43. Frontispiece of *A Modest Enquiry into the*      196
    *Nature of Witchcraft*, 1697

## LIST OF MAPS

1. The Modern Global Distribution of *Ixodes* Ticks That      19
   Can Transmit *Borrelia* infections
2. North America at End of Last Glaciation, 18,000 Years B.P.      38
3. Lyme Disease in Massachusetts, 2002      79
4. Lyme Disease Risk Map: U.S.A.      86
5. The English Colonies: Witchcraft Accusations      87
6. Map of the Lyme Risk in Defined Regions in Europe      88
7. Areas of Witch Hunts in Europe: 1400-1800      89
8. Territories in Europe belonging to Spain in the sixteenth Century      90
9. Salem Area Growth: 1628-1692      181
10. Pre-1628 Native American Settlement and Land Use      182
11. 17th C. Landscape Features      183

INTRODUCTION

# LYME DISEASE: A PAST AND PRESENT PROBLEM

This work began when I read for the first time a written list of the symptoms of some of the 'afflicted' from the Salem Witch Trials of 1692. I found some of the experiences to be startlingly familiar as symptoms that I had personally experienced while suffering from Lyme disease. A quick perusal of historical information about witchcraft accusations and witch-hunts in the past revealed an interesting overlap in symptoms, geography, environmental conditions, and demographics between modern Lyme disease and historic witch-hunts. Any study of European witchcraft quickly reveals that both witches and many of their supposed victims had distinctive red marks on their skins. They were described as suffering from a variety of neurological and arthritic conditions, in addition to the more fantastical beliefs associated with witchcraft, like flying through the air on broomsticks or creating spectral hauntings. This led me on a long journey of discovery that has ended with a belief that Lyme disease afflicted people in the past in both North America and Europe and was associated with the persecution of witches in the early modern age. The same European areas that were rift with witchcraft accusations in the past are endemic for Lyme disease in the present, as is Massachusetts, where the town of Danvers was once known as Salem Village.

The hypothesis of this book is a simple one: that disease has shaped culture and culture has defined disease throughout history. This disease hypothesis, however, depends upon identifying the proper diseases. The people that live within our modern societies are all the descendants of ancestors who suffered from diseases in the past. Deadly diseases that were faced not only fill our historic cemeteries, but have threads interwoven into the fabric of our cultural memory and our biology. Humans have been confronted by several disease driven apocalypses in the near past. Disease has shaped our history. Our nursery

## DISGUISED AS THE DEVIL

rhymes warn that "pocket full of posies" will not save us because eventually "we all fall down" and that opening a window may bring "influenza." Smallpox killed, was behind the discovery of a vaccination and an eighteenth century fad for decorative scar concealing face patches. Less deadly diseases, however, may have a greater cultural effect than ones that kill and then depart. A less deadly pathogen, like Lyme disease, has a steady accumulative effect: the disease affects the culture and the culture affects the disease through time in an ongoing systemic cycle. These effects can become so intertwined and interwoven that they reach a point of near invisibility. Social definitions come and go, while the evolving natural environment provides a backdrop, and human nature seems to remain fairly steady. So, while neighbors have probably been squabbling since the Stone Age and continue to in the present day, witchcraft accusations have been recorded only at certain points of historical time. For a witch to exist it must first be culturally constructed.

The Early Modern Era's witch-hunts that occurred in both Europe and North America appear to follow a very specific script. The conditions that must be met for witches to appear are related to: 1. CULTURE- the culture must not only understand but believe in the existence of witches, 2. PLACE- the population inhabits a localized rural/suburban geographical area that is located in the northern hemisphere, usually around or above the 40 degree latitude mark, 3. DISEASE - both the accused witches and their implied victims are afflicted in a culturally constructed way, often by the interpretation of a specific set of physical and neurological symptoms, and some will have distinct red marks on their skin, and 4.THE ENVIRONMENT[1]- specific environmental conditions must exist, the weather may be poor, and people are either chopping down the forest in a splotchy pattern or the forest has re-grown. Once people live on land that is cleared of trees, are surrounded by their environmentally modifying domesticated animals (especially their pigs) and achieve social stability-witchcraft accusations start to evaporate.

Neighbors continued to squabble and people may still sporadically become sick, but as we move through time from the past towards the present day, the scientific paradigm of the Enlightenment, which does not acknowledge magic, becomes predominant. Witchly malfeasance is no longer a strong cultural belief, although a full set of folk superstitions and 'knowledge" is stubbornly retained, and, indeed appears in the modern day. We revel on Halloween. A cardboard witch has a wart or mark on her face in addition to her pointy hat and broom for flying. Everyone knows that a black cat crossing a path is somehow unlucky even if we

## INTRODUCTION

may not know why. Ghost hunters appear on cable television. The modern Wicca movement has many converts. And, just when the forests of New England and Europe had grown back, it was cut down once more by suburbia in a sprawling splotchy pattern and the affliction that we call Lyme disease reappeared. Just as there was a politics of witchcraft accusation and legal justice in the past, there is a modern American politics of Lyme disease in the present. Public policies like federal highway construction and the mortgage availability of the post WWII G.I. Bill were used to create this landscape of sprawl that is associated with this disease. Lyme disease research itself is very much a work in progress that has sometimes raised more questions than it has answered. Lyme disease appears to be a modern "devil in disguise."

Any study of witchcraft is potent with the possibility of controversy but I was surprised to find that Lyme disease is also a 'lightening rod' topic. My interest in this subject is openly motivated first by my background as a patient and then by my interest in the experiences of other patients. Over the past ten years, I have come into contact with many members of the medical profession and my contemporary Lyme sufferers. Through this research I feel that I have also found sufferers in the historic past, including an ancestor who was accused of being a witch in the 1692 "witch craze" and died in prison. After finishing this research, I am left wondering whether I inherited the HLA-DR4 or HLA-DR2 genetic trait from her that made me more susceptible to this disease along with her red hair. The story that can be told is a compelling one.

I began this research with the idea of telling the story of Lyme disease from a historian's point of view. I would use the very best of available modern research and apply these scientific principals to the past; this would be in the form of an interdisciplinary approach in a form could be called enviro-history. What I found surprised me. Modern American Lyme disease research is a very complex tableau of contrasts: innovative science sharply framed by controversy and complaint set against a dark backdrop of conflicts of interest and questionable medical ethics. It is difficult to know what research to use to apply to the past. In the end, I tried to use it all. Sometimes modern knowledge is a battle royal. Some of the most shining Lyme-epidemiological heroes of the late nineteen seventies have found their reputations to be somewhat tarnished in some circles in more recent times. One of the Yale researchers who helped give what he considered a new disease its name now works in a laboratory that has to be protected by guards. He has received death threats. When he publicly lectures on his pioneering work, he has been

## DISGUISED AS THE DEVIL

picketed by members of the Lyme Activist and other patient advocacy groups.[2] This group objects to the doctor's decision to declare that Lyme disease is over-diagnosed and to back away from his own initial definition of a persistent third, or chronic, stage for Lyme infection, identifying it instead as some sort of post-Lyme syndrome or fibromyalgia. This redefinition helped limit the length of antibiotic therapy that many medical insurance companies will pay for, and may have given rise to a surge in the diagnosis of fibromyalgia as yet another rapidly emerging condition during the late nineteen eighties.[3] The topic of antibiotic usage is a hot button issue. Because of the siloed nature of modern medicine, for non-dermatological patients antibiotics came into the picture fairly late and, some would argue, with too little usage. When Lyme disease was first '*named*' by American medical doctors, the causative agent was unknown. It could have been either viral or bacterial or anything else.

It was several years before Dr. W. Burgdorfer discovered the Lyme spirochete *Borrelia burgdorferi* (spirochete is the class, *Borrelia* is the genus, and the species of any germ is often named after its discoverer). It may have taken too long for this discovery to be accepted within the medical community because antibiotic usage was a hit or miss thing until the mid-nineteen eighties. The same doctor appears to have continued to study the long-term outcome of this disease without antibiotics, even after he had published an

Figure 1. *Fairfield County Weekly, Connecticut, May 20, 1999*  article showing that antibiotics could be effective in resolving its symptoms. For the historian these antibiotic-less studies provide excellent guidance for symptoms that might be described in historic records. For the modern ethicist the

## INTRODUCTION

delay in treatments may be troubling. Some of these early studies bring the dark specter of the Tuskegee experiments into the world of Lyme research. One scientific inquiry, for example, looked at Russian patients who developed the classic Erythema Chronicum Migrans (hereafter ECM but sometimes also called an Erythema Migrans or EM) bull's eye rash of Lyme borreliosis. These patients were observed but *not treated* at a time when it seems to have been *known* that antibiotics would deter the development of more devastating symptoms. Such a study would have difficulty winning approval from an institutional research oversight committee today. The morality of this study can be questioned even if it occurred on foreign soil in a country with a political system and history that is very different from our own.[4]

    The scientific literature itself is fraught with contradictions. There is a stark dichotomy of opinion about the severity of Lyme disease. In many people, it is a minor affliction that is easily treated. Clearly, Lyme disease is not life threatening in the vast majority of cases. However, numerous articles persisted in repeating the mantra that "no one has ever died from Lyme disease."[5] A clear record of deaths directly contradicts this statement. Since many autopsies do not routinely include tissue and fluid testing by the silver or immunological staining needed to find the Lyme disease spirochete, the death rate is currently unknown.[6] Lyme disease is described as both "an easily treatable condition" and a stealthy "complex, difficult to diagnose" problem that is "the leading cause of vector borne illness in the United States today."[7] Most disturbing may be the level of disability that has gone unrecognized because of this conflict.

    The two key blood tests, the WESTERN BLOT and the ENZYME LINKED IMMUNOSORBENT ASSAY, or ELISA, that are currently used to diagnose the disease are also subject to controversy. They are notorious for false results. When the United States Congress passed Public Law 107-116 it noted that "the current state of laboratory testing for Lyme disease is very poor."[8] The situation has led many people to be misdiagnosed and delayed proper treatment. The Lymerix vaccine clinical trial study has documented that more that one-third (36 percent) of people with symptomatic Lyme disease may not test positive on even the most sophisticated tests available. The study also proved that Lyme disease is hyper-endemic, and seriously underreported. The group studied showed per capita infection rates that exceeded 1000 per 100,000, both in year 1 and year 2 of the study, even following vaccination! The ramifications of these deficits of knowledge in terms of **unnecessary pain, suffering and cost are staggering.**[9] [When President George W. Bush signed this bill

into law, little did he know that within five years he would become one of this affliction's most famous victims] Even the very determination of what constitutes a positive test result changed in 1994 when the U. S. Food and Drug Administration (hereafter FDA) removed two of the most important outer surface protein (Osp) bands from test results as positive diagnostic markers. This was to prevent false positives due to their use in vaccines. Some Lyme activists see a conspiracy in this action, saying that Lyme disease itself was redefined to facilitate the development of lucrative vaccines. Lyme activists have protested that this new standard effectively eliminates many of those who are infected with the Lyme bacteria from ever testing positive.

**It has been suggested that this new standard was set to make an ineffective vaccine (that both the federal government and a drug company had financial investments in)** appear to be highly successful by seeming to protect a large percentage of a population that would never ever again test positively. There appear to be some conflicts of interest between those who stand to profit from the vaccine: university patent holders, drug companies, and the FDA itself as another patent holder-and the welfare and health of the American public.[10] The vaccine made some apparently healthy people who participated in testing trials sick, and sickened more after it was approved for general use on adults in endemic areas. By the time the vaccine was pulled off the market, reported side effects from the LYMERIX vaccine included 640 emergency room visits, 34 life threatening reactions, 77 hospitalizations, 198 disabilities and six deaths.[11] These maimed victims seem to either have had a pre-existing infection that the vaccine reactivated or a genetic sensitivity related to HLA-DR4 and HLA-DR2 alleles that caused their immune system to go into hyper-drive and attack their own body. This resulted in some permanent arthritic conditions. This particular genetic trait is shared by about thirty percent of the modern American population. It may be more prevalent in modern England.[12] Studies have also shown that systemic exposure to OSP's causes immune suppression which allows for the activation of latent viruses of all kinds (Epstein-Barr, Cytomegalovirus, HHV-6, etc.) and funguses that would otherwise be controlled, to multiply and thrive in some individuals, causing chronic debilitating conditions.[13]

When the Smith Kline Glaxo Company voluntarily recalled the vaccine due to "a lack of interest," it is estimated that they had already reaped a multimillion-dollar profit from it. The class action lawsuit related to this vaccine that was settled in 2003 seems to have barely

## INTRODUCTION

dented the drug industry's enthusiasm for Lyme vaccine profits. Lyme disease vaccine research is also still heavily funded by the National Institute of Health, the National Institute of Allergy and Infectious Diseases, and the FDA. [14]

The public, however, has become wary of these vaccines which may reduce future sales. Nevertheless, after trying human vaccines, the drug industry has now turned their attention to pets and wildlife, especially after animal pharmaceuticals were placed under the auspices of the FDA regulatory system for the first time. A vaccine for dogs has been available for several years. A recently published study has suggesting that we could control Lyme disease by vaccinating all the wild mice in the United States.[15] Since urban rats have also been found to be extremely competent vector hosts for the Lyme bacteria, they may soon become potential vaccine recipients.[16] Shrews, chipmunks, and other wildlife species have also been found to be vectors for this bacteria so the list of animals that could be lined up for annual vaccine booster shots is a long one. [17]Wildlife have the added benefit for drug producers of not being able to sue anyone if they are adversely affected! The profit from such a plan could become a bonanza of historic precedence. The fact that laboratory rodents have developed arthritic joints when infected with the *Bb* bacteria or Osp A vaccine has not dampened the enthusiasm for this program.[18]

As of today, federally funded Lyme disease research has gleaned over a billion dollars in grants, and worldwide sales of testing supplies for this single disease could annually top the billion-dollar mark within a short while.[19]This tidal wash of cash has found neither a gold standard diagnostic test nor created scientific agreement. It seems to have done little to halt the continuing spread of Lyme disease in America or protect the population from the disease. The United States Center for Disease Control reported that the reported incidence rate for Lyme disease jumped by forty percent between 2001 and 2002, and is currently at an all time high.[20] It continues to climb.

The duration of treatment for Lyme disease is a contentious issue. There are currently two major medical associations that have issued conflicting guidelines for the diagnosis and treatment of Lyme disease: the Infectious Diseases Society of America (IDSA) and the International Lyme and Associated Diseases Society (ILADS). These "two standards of care" fuel another ongoing controversy. Lyme support groups often recommend the use of "Lyme literate physicians" who provide treatment using the ILIADS guidelines, which may include lengthy antibiotic usage

 DISGUISED AS THE DEVIL

for chronic cases. These physicians, however, have become vulnerable to Medical Board Reviews in many states. Other professionals follow the IDSA guidelines, which have effectively removed third stage or chronic Lyme disease along with any treatment from the discussion.

A primary factor in this controversy is that the disease can be difficult to diagnose. Not every patient suffers from the typical "bull's eye" ECM rash or the joint inflammation that are considered classic symptoms of Lyme disease. Only fifty to sixty percent of patients typically recall a tick bite; the ECM rash is reported in thirty five to seventy percent of patients (depending on which study you read) and joint swelling typically occurs in only twenty to thirty percent of patients. Given the prevalence of modern over-the-counter anti-inflammatory medications like Ibuprofen, joint inflammation may often be masked. Some patients with Lyme disease will continue to experience a variety of symptoms long after initial treatment. Some of these patients go on to develop the prolonged multiple nonspecific symptoms that are sometimes called chronic Lyme disease. Studies have shown that *Borrelia burgdorferi* can persist long after antibiotic treatment. Studies conducted in animals - including mice, dogs and monkeys - indicate that the bacteria can persist after treatment is thought to be complete. Persistence in humans has been confirmed by culture or molecular testing in at least a dozen studies but the science in this area is still evolving.[21]

Protection is another issue. The common sense list of personal precautions to protect against ticks has not changed in the last fifteen years (see figure 2) but has never actually been subjected to scientific blind tests to verify efficacy.[22] This may be due to an almost complete lack of funding for any disease control method that will not yield profits for some corporation. The recommended precautions are difficult to follow in the heat of summer and may have to be forcibly imposed before they could ever become effective. Soldiers, when required under threat of court marshal to wear uniforms and tuck their trousers into their socks in the summer heat, for example, *do* test as significantly less likely to be seropositive for either Lyme antibodies or anti tick saliva antibodies (ATSA) when compared to normally attired control groups.[23]

Chemical tick repellants are recommended. However, the continuous use of the most recommended tick repellents containing N, N-diethyl-meta-touamide, or DEET, may also be problematic. DEET has an interesting history. It was first developed for use by the U.S. Army in 1946, and is a byproduct of the Cold War Era's bio-weapons programs.

# INTRODUCTION

# It's Lyme Time!
## Protect Yourself Against Lyme Disease*
## in Spring, Summer, and Fall

**1** Walk in the middle of trails, away from tall grass and bushes.

**2** Wear a long-sleeved shirt.

**3** Wear white or light-colored clothing to make it easier to see ticks.

**4** Wear a hat.

**5** Spray tick repellent on clothes and shoes before entering woods.

**6** Wear long pants tucked into high socks.

**7** Wear shoes—no bare feet or sandals.

*Lyme disease, the most common tick-borne disorder in the U.S., can affect the skin, joints, nervous system, heart, and eyes.*

**Figure 2.** *Recommended precautions for Lyme Endemic Areas: Preventionguy from www.lymediseaseassociation.org.*

## DISGUISED AS THE DEVIL

Various species of ticks and insects have been experimentally infected with a variety of pathogens during the last century. Ticks were infected or weaponized by both the Nazis and the former Soviet Union. DEET was developed to protect American military personnel against this threat. The official EPA information sheet for DEET states that it has been found to occasionally cause seizures in small children.[24]

Current treatment protocols for Lyme disease range from those described as "highly effective" to the standard treatment for third stage Lyme disease being described by one activist as "no treatment at all." Lyme disease science is described as being both cutting edge and unresponsive to the afflictions of patients. There seems to be a sharp dividing line between the Lyme researchers who develop theories, consult for insurance companies, and have monopolized research monies for the past several decades and the medical practitioners who are in the field dealing with the Lyme afflicted on a daily basis. Patients often feel lost in a medical morass, stretched to the limits of their pocketbooks, with their cultural expectations that science will provide solutions for their health problems shattered. [25]

Culture, as used here to define expectations, refers to the language, accumulated knowledge, beliefs, assumptions, and values that are passed between individuals or groups over the generations. A culture has a system of meanings and symbols that shape how people see the world and their place in it. It gives definition to personal and collective experience. Modern American culture is strongly rooted in the Enlightenment's belief in the supremacy of the scientific method and science itself. Lyme disease may yet test this modern cultural belief in science to its limits.

Lyme disease appears to be an ancient and ubiquitous affliction that humans have been coping with for eons in areas that are infested with ticks of the *Ixodes* species. Using the tools available to them within their own particular culture or social paradigm, they labeled and understood the disease as a variety of named afflictions. These may have included witchcraft affliction, rheumatism, scurvy, the summer sickness, neurasthenia, and currently fibromyalgia. When I went looking for Lyme disease in the pre-enlightenment time period, I kept finding witches. Witchcraft accusations and afflictions appear in the historic records during distinct time periods, usually related to the expansion or contraction of oak forests, in almost the identical geographic areas where people are plagued with Lyme disease today.

# INTRODUCTION

*How would people have fared if they confronted an affliction like Lyme disease in the past?* Using the European description of witchcraft affliction as a set of very similar physical symptoms [surely as baffling a set of symptoms as has ever been described in any culture] as delineated in witchcraft texts as an example; the answer seems to be **not too well**. The people who populate the historic records were treated in a variety of ways by their medical and political establishments and popular cultures. In the same time period some people, with very distinct characteristics, were labeled as witches while others were called their victims. Often neurological symptoms were involved and for the historian the distinction between the two groups seems blurred.

During the events in New England that culminated in 1692, a dichotomy emerged in how the physical afflictions of witchcraft were considered. When there was public support for the diagnosis of witchcraft, the Massachusetts Bay Colony threw massive amounts of medical and political attention and legal power at the problem. When public opinion questioned this diagnosis and the public reputations of some of the public figures involved were at stake, the symptoms of affliction became less important. In Boston, for example, the fits and seizures of a young woman that occurred shortly after the famous 1692 trials were over became almost a form of public entertainment-a seventeenth century precedent for reality television. A crowd would often stop to gawk at her 'afflictions' on their way home from church. And time has not been particularly kind to some of the afflicted from 1692. The accused witches of New England now garner most of history's compassion.

Partially because of their own later behavior, and especially after the publication in 1700 of Robert Calef's acridly skeptical *More Wonders of the Invisible World*, the afflicted became immortalized as "a parcel of possessed, distracted or lying wenches" whose morality was that of "vile varlets." When Thomas Hutchinson wrote about the subject in 1750, he concluded that the events at Salem had been "fraud and imposture" even if "a great number of persons" thought the afflicted had "been under bodily disorders which affected their imaginations." Two of the afflicted girls seem to have contributed to these pronouncements by their later actions. Suffering terribly from affliction as a girl in 1692, Mercy Lewis played a prominent role in one of Cotton Mather's early essays on witchcraft. She later gave birth to a child out of wedlock, which created a scandal in Puritan circles. In 1698 another afflicted girl of 1692, the by then married Mercy Short Marshall, was found guilty of adultery and

 DISGUISED AS THE DEVIL

excommunicated from the church by her pastor, Cotton Mather, who had earlier been one of her staunchest supporters.[26]

Very little is known about the later life of other of the New England "afflicted." Abigail Williams is supposed to have remained sick and died young, probably sometime before 1697. She was described as being "followed by diabolical manifestations to her death." We know that the afflicted Betty Parris grew up, married, and lived to an old age in relative obscurity. [27] Elizabeth Knapp, a possessed girl from Connecticut is known to have fully recovered, married Samuel Scripture in 1674, and had six children and a marriage that lasted at least twenty-five years. Other than a comment about her gaining weight, her name disappears from the record books.[28] The Goodwin children who had suffered affliction in 1688 apparently "returned to their ordinary behavior, lived to adult age, made profession of religion, and the affliction they had been under they publicly declared to be motive to it." One of the Goodwin girls in later life "had the character of a very sober virtuous woman, and never made any acknowledgment of fraud in this transaction."[29]

By focusing on the social aspects of the New England incidents, the deeds of *Wayward Puritans* and *Salem Possessed*, the actual symptoms of affliction that were witnessed by some of the most prominent citizens of the Massachusetts Bay Colony have often been overlooked, minimized, or decried as fraudulent. The sense of urgency created by the sudden onset of illness and ongoing affliction has sometimes gotten lost in the shuffle of scholarship. Did the afflictions continue even after several people were executed for causing them? The answer seems to be yes. For Abigail Williams the affliction was soon fatal. Others lived on and continued to suffer from what could be called chronic affliction. When Ann Putnam, Jr. stood up in church several years later to apologize for her earlier actions, she stated that she had long "continued to be ill." Her mother, who had also been afflicted in 1692, continued to suffer symptoms afterwards and died within a few years of the witchcraft trials.[30]

To the modern mind, the concept of witchcraft has lost its cultural potency while retaining a riveting sense of entertainment and fascination. Americans have been terrified by the Blair Witch Project and entertained by the world of Harry Potter. While 1692 stands out in our collective history because of the written record, the events of that year may be part of an ongoing American saga that ends with the Lyme afflicted of today and starts at the moment of New World contact by humans with a tick risky environment.

# INTRODUCTION

The physical description of those who were accused of being witches in 1692 is extremely sketchy. Keeping in mind the prevailing life expectancies at the time, some were older women, some were ill and afflicted themselves, and many were known for having temperamental dispositions as argumentative scolds who might be found muttering to themselves and arguing with family, friends and neighbors. Some were among the "usual suspects" with a rap sheet of prior witchcraft accusations as part of their life history. At least one, according to oral tradition, had been red haired in her youth.[31]

All were subjected to a full body search for a 'witch's teat' or 'devils mark' on their skin. Some had animals like dogs or cats that lived with them. In the seventeenth century, these animals were suspected of being the 'familiars' that were used by witches as evil messengers. In seventeenth century culture, the concept of a house pet was unknown or considered strange. Domesticated pets are known to be the carriers of a variety of ticks and may make their owners slightly more likely to contract Lyme disease.[32]

Lyme disease also has an environmental context. In many ways, Lyme disease assaults modern cultural biases because we no longer have the seventeenth century's dread of a devil-inhabited wilderness. Instead, the wilderness has become part of Mother Nature. Lyme disease is contracted when we contact nature, which is supposed to be, as Martha Stewart would say, "a very good thing." As a culture, we have preserved wilderness areas and gone back to nature. We eat natural foods. We commune with nature. We contribute to the Fresh Air Fund. We rally on Earth Day. We support environmental causes. It is almost unfathomable that such a good thing could make anyone so sick. So we equivocate.

In the nineteen eighties Lyme disease was part of a 'March of Progress' mentality with a high profile as a *new* disease uncovered by diligent American scientists soon to fall victim to the best scientific practices available. When modern science had not eradicated it by the nineteen nineties, it moved to the edges of the national radar screen and stays there as more and more people become infected. The disease's very complexity may be more than the American public is willing to comprehend.

The modern debate about the affliction of Lyme disease is complicated by the ongoing controversy about whether this is in fact a *new* disease, perhaps the result of some errant escape of germs from the United States Department of Agriculture research lab located off the coast of Long Island, New York, at Plum Island, [33] or a very ancient form of

## DISGUISED AS THE DEVIL

affliction that peaks and ebbs in response to changing environmental conditions.³⁴ It may be both, but this work will investigate the concept of Lyme disease as an *old* infection. With a few stops in Europe for background information, it will focus on New England, because that area has high modern endemic levels of infection, which may have an antecedent in the seventeenth-century, and there is an extensive documentary record where sporadic afflictions are recorded but given diagnoses that are idiosyncratic to a specific time or century.

In his book *Making Sense of Illness*, Robert Aronowitz discusses the processes by which we recognize, name, classify, and find meaning in illness. He specifically discusses Lyme disease in a chapter entitled *"The Social Construction of a New Disease and Its Social Consequences."* The social construction of a disease is the way that non-biological factors like beliefs, economic relationships, and societal institutions greatly influence, if not define, a cultural group's understanding of any particular illness.

Aronowitz uses the modern "discovery" of Lyme disease in the early 1970's as a classic case study for the social construction of disease. Lyme disease was conceptualized as a *new* disease for a variety of reasons that influenced the prevailing interpretation of a set of biological and epidemiological facts. If the outbreak had occurred in Europe, it would have been diagnosed in a very different way based upon past practice. The decision that this was a newly emergent disease was influenced by the nationality of the investigators (American vs. European), the disciplinary background of the medical researchers at Yale University who named the disease (rheumatologists vs. dermatologists), intellectual or attitudinal features (American skepticism of the quality of past research and European intelligence) and professional concerns (potential self-interest in being the medical pioneer who discovered a new disease).

European dermatological research, dating back to as early as 1910, had described an expanding ring shaped rash that developed at the site of a tick bite. Studies done in Norway during the 1930's identified possible bacterial spirochetes in these lesions. A study done in 1955 found penicillin to be highly effective for treating tick bite related rashes. In 1970, North American dermatology textbooks described this rash as part of an infective process that resulted from a tick bite, was possibly caused by spirochetes, and was responsive to antibiotics.

The first modern Center for Disease Control (CDC) reported case of an ECM in the United States was from Wisconsin in 1969. The first modern case cluster of Lyme disease in the United States was identified

## INTRODUCTION

by dermatologists working at the Naval Submarine Medical Center in Groton, Connecticut, during the summer of 1975. It was diagnosed as an outbreak of ECM and effectively treated with antibiotics.

It was the rheumatologists from Yale University, including Dr. A. Steere, however, who were called in to study a group of sick children living in Lyme, Connecticut that would shape America's perceptions. Unlike the dermatologists, they dubbed the cluster of sickness they found there as something new. It was called Lyme arthritis. Severed from its historical past, the disease went on to be publicly accepted in the United States as a newly emerging disease that was 'discovered' by American medical researchers.

Over the next decade a myriad of non-arthritic symptoms that were caused by the same bacteria began to be recognized including among others: neurological problems, the set of symptoms called Bannworth's syndrome, and two other types of skin lesions that are found in Europe: Acrodermatitis Chronica Atrophicans and Borrelial lymphocytoma (a raised red bump). It became a confusing situation. At the behest of Dr. W. Burgdorfer, the discoverer of the *Borrelia burgdorferi* spirochete (and a Lyme victim himself) the full set of afflictions was given the uniform name of Lyme disease. The Erythema Chronium Migrans rash, the European name of the disease in and of itself, was relegated to become the name of the ailment's characteristic *bull's eye* rash.[35]

By conceptualizing Lyme disease as a new disease, the need to look for its role in the past was essentially eliminated. When it *is* looked for, there is evidence that challenges this perception of newness. For example, DNA amplification work has found that the *Borrelia* bacterium is of ancient origin, dating back at least to the Paleolithic Age if not earlier and probably of European origin.[36] The complex shape shifting stealthy attributes of the *Borrelia burgdorferi* [hereafter *Bb*] spirochete is another clue that it is the result of a lengthy evolutionary process. *Bb* and other bacteria, the *Ixodes* family of ticks, and other vector hosts have been coexisting for a long time in many parts of the world with populations that have expanded, contracted, and changed in response to ecological and other forces.

The advance and retreat of glaciers, reforestation, deforestation, and changing animal host population growth, including that of humans, are all important factors in the history of this disease. Part of the ecological process for Lyme disease may have even been understood by early Native American groups who linked fever and joint pain with human interaction with deer. They constructed the sickness as being the direct result of

## DISGUISED AS THE DEVIL

disrespect for the spirit of the deer, cloaking it in religious overtones but missing the speck-like ticks that were the true purveyors of misery.

When the earliest English colonists and their domesticated animals began to occupy forest edge environments in Virginia and parts of New England, they also suffered from ailments at "contact" that are compatible with a diagnosis of Lyme disease. These early settlers constructed their definition for their sickness by contrasting their existence with the life they had left behind in England. They were sick, they felt, because their new life was filled with *starving times*, exposure to the elements, and scurvy. The water they were forced to drink (at least in Virginia) was inferior to that which was found in England. They believed that their cows languished because they were forced to eat American grass that was either too rich or too poor when compared to their accustomed diet of good English grass. To some of these settlers the grass was truly greener back at home on the other side of the ocean.

By 1692, a long list of afflicted persons (and their animals) had sporadically experienced symptoms for more than eighty years. The Salem Witch Trials gave some of these people that history would usually ignore the chance to describe their personal experiences for the written record. The Lyme etiology of tick attachment, rash, and symptoms is described in terms that are idiosyncratic to the seventeenth century. Testimony records the preternatural appearance of pins sticking into the flesh of victims, various skin rashes that are often described as bite marks, lethargy, lameness, and swellings, and a long list of neurological afflictions and hallucinations.

The landscape that these victims and accused witches lived in, interacted with, and described, when plotted onto a map, contains large areas of riparian and fragmented forest edge terrain that would be classified in modern times as tick-risky. Young children and women who traversed this landscape in long (tick collecting) skirts and petticoats seem to have suffered from dramatically higher rates of infection than their male counterparts, who wore tucked in clothing that was more protective. Mysterious illnesses also struck domesticated animals.

There is a long list of Lyme-like neurological symptoms that are described in the seventeenth century record: it includes a child with the symptoms of Bell's palsy, children with seizures, the sensation of heat and pricking on the skin, transitory sensory loss, irritability, lethargy, and occasionally a fatality. Some of the sicknesses and afflictions of the seventeenth century were medically, socially and culturally constructed as being the result of the curse of a witch. Because they fit a specific set of

# INTRODUCTION

symptoms that had become well defined in the folkloric knowledge and written texts of Europe, the cultural and social acceptance of this definition led to the legal prosecution of those accused of practicing witchcraft.

The English cultural, folkloric, and medical practices in use during the seventeenth century offer some insight into the possible origins of various afflictions. Folkloric 'cures' for scurvy (a disease now defined as a deficiency of vitamin C), for example, included juice from lemons, rose hip tea, and herbs that were rich in vitamin C. But they also included remedies that have no vitamin C at all but do have antibiotic properties. The origin of the affliction of scurvy in England, based on the various traditional 'cures' used repetitively,[37] point to both bacterial infection and vitamin C deficiencies or an interplay between the two, as causative. Lyme disease may have been known and treated in Elizabethan England as a disease called 'land scurvy', which was prevalent in the spring and summer. On the other hand, many of the folkloric herbs associated with use by witches and the devil were properly feared as evil-they are deadly poisons.

Some of the herbs and substances that had a folkloric reputation for protection from witches are either insecticides or insect and arachnid repellants. Seeking protection from witchcraft, the afflicted Mary Hortado[38] hung bay leaves, which have been found to repel insects, on the door of her seventeenth century home in Maine. Ivy was also used for protection from witches. It repels insects. A quaint English cottage with ivy-covered walls was protected from both witch and insect. This may indicate a connection between the afflictions thought to be caused by witches and a variety of insect and arthropod related diseases.[39]

Over the course of one century, Lyme disease may have gone from being defined by the Native American occupants of North America as an illness associated with deer, to being scurvy at Plymouth and Massachusetts Bay, to being the result of witchcraft. By the twenty-first century, it is once again being defined as a disease associated with deer. Lyme disease may also have acquired some other labels during its passage through time. Some of the upper class "rusticators" of the early twentieth century, for example, after reestablishing contact with a reforested but *healthy* natural world suffered from neurasthenia,[40] an ill defined disease which had some very Lyme disease like neurological and fatigue characteristics. Nineteenth century culture had little patience for this or any other mental affliction, especially when it was suffered by hysterical females. Writing in 1895, Sigmund Freud commented that "during the last few decades a

 DISGUISED AS THE DEVIL

hysterical woman would have been almost as certain to be treated as a malingerer, as in earlier centuries she would have been certain to be judged and condemned as a witch or as possessed of the devil."[41] Modern physicians have depicted patients who experience the symptoms of neuroborreliosis as having the class of disease popular in the Post Freud Era: the psychiatric disorder. This disorder is indeed an omnipresent affliction in most human societies. A recent study has shown that as many as one third of all the patients in one modern psychiatric institution may be infected with the *Bb* spirochete.[42] When the twentieth-century patient, Polly Murray (who would later write a book about her experiences) was admitted to the mental ward of a Boston hospital,[43] she may have been reluctantly following in the footsteps of many centuries' worth of Lyme disease sufferers. Lyme disease, called "the great imitator" because it carries with it such a wide array of symptoms,[44] may have been an omnipresent force throughout history.

This study will explore the environmental and cultural context for Lyme-like disease in the past, assess risk, and attempt to contrast the personal perceptions in the historic records with modern research. Using scientific information available from modern sources, it may be possible to find evidence for Lyme disease lurking in historic books and documents, hidden by the haze of time. For example, if it was June of 1630 and if you had just taken a very long sea voyage and you came down with a fever, your joints began to ache, and you were too tired to move- even though you were eating strawberries-was it unreasonable for you to "know" that you suffered from scurvy? We now know that strawberries are packed with vitamin C, making scurvy an unlikely diagnosis. But 'land scurvy' may be another matter. We also know that June is a month where there is a high risk for contact with nymphal *Ixodes* ticks and the bacteria that they may carry in Massachusetts. It is therefore possible to suggest the plausible alternative diagnosis of Lyme disease (then being called land scurvy) for some of the afflicted.

By examining descriptions of personal experiences written during the seventeenth century, an understanding of how the first settlers in Massachusetts viewed and interacted with their environment can also be formulated. It may provide valuable insights into the past that might also be applicable to the present. The story of the history of Lyme borreliosis as an *old* disease would read something like the following. After the retreat of the ice at the end of the last period of glaciation, the dynamics of climate, trees, mast [acorns and other tree nuts], deer, mice, birds,

# INTRODUCTION

ticks, and bacteria began a long and intrinsically complicated set of interactions that still occur in modern times.

For the most part, they take place across a large swath of the earth's land surface- girdling the forty to fifty degree latitude marks in the northern hemisphere, wherever *Ixodes* ticks are known to thrive. Today this swath may be expanding due to global warming. A report from Sweden notes that these *Ixodes* ticks are now found near the Arctic

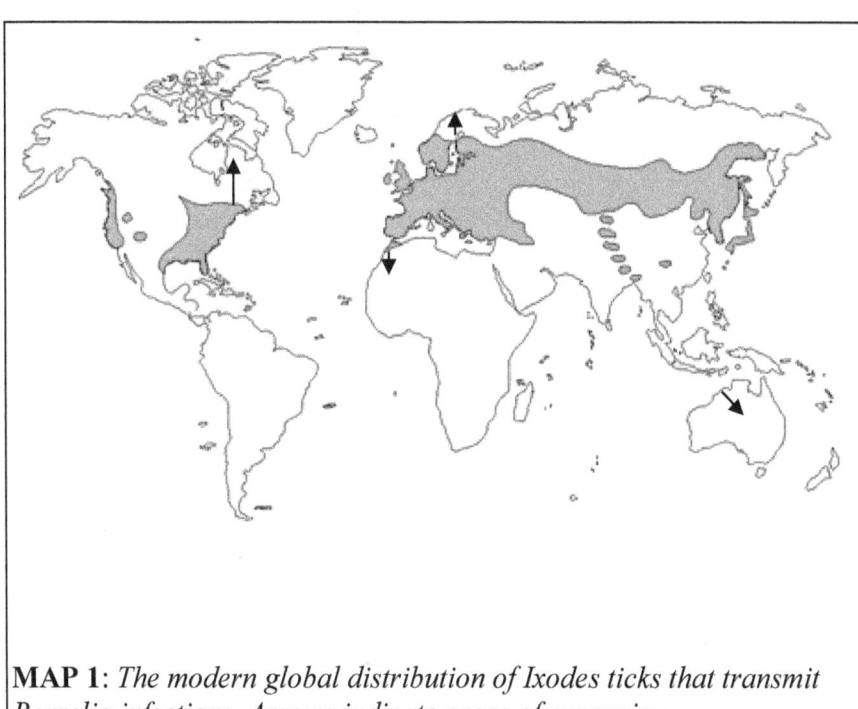

**MAP 1**: *The modern global distribution of Ixodes ticks that transmit Borrelia infections. Arrows indicate areas of expansion.*

NWDIII

Circle. Another study suggests a rapid tick and rodent host movement in northern Africa due to the climate change related drought. When environmental conditions are just right, Lyme disease may become endemic, often in a dispersed, splotchy, localized, dynamic with edges that are also blurred by the movement of various species of animals and birds.[45]

In postglacial North America, as the ice melted and forests re-grew, *Ixodes* ticks were carried further and further north as their blood meal hosts moved into re-expanding beech and then oak dominated

deciduous forests.[46] This new sparse tick population had fewer choices for mating and eventually became less genetically diverse than the relatives they had left behind in the south. This made these northern ticks into an even more compatible host for the bacteria evolving within their guts-the spirochete that would become the modern *Bb,* along with other germs capable of causing a variety of symptoms, including *Anaplasma (Erhlichia), Babesiosis, Bartonella, Mycoplasma* and *Rickettsiae-* the classic co-infectors of Lyme disease. These *Ixodes* ticks fed on a variety of blood meal hosts. Some hosts, like lizards in southern areas and white tailed deer everywhere, have been coexisting with *Bb* for so long that they evolved to acquire natural immunities to these bacteria- substances within their blood attack the Bb infection before it can spread. Other species, like white footed mice, have body systems that can be infected with bacteria without exhibiting many life-threatening symptoms. It may be that many species have acquired, over time, the ability to harbor the *Bb* within their bodies without becoming sick or with only low level or almost unrecognizable effects. This makes *Bb* a very efficient pathogen-it survives without sickening or killing its host.

And it, in turn, has developed a full set of survival skills to keep from being destroyed by any host's immune system. When its movements were recently clocked by scientists they were found to be the fastest ever recorded for a spirochete, moving at a rate that is two orders of magnitude speedier than the fastest moving human body cell. This "alacrity" in an organism that has bidirectional motor capacity, the scientists concluded, may well contribute to difficulties in spirochete clearance in a host. Cells involved in the human immune system may not be fast enough to grab it![47]

In addition to speed, the bacterium is a shape shifter. Drop a Lyme spirochete into a vial of distilled water and it will transform itself. The spirochete will develop a tough cyst to encase and protect the vital genetic material within its structure that it needs to replicate. Inside the blood stream of a host animal the spirochete may sacrifice bits of its outer casing proteins- spitting them off in a process called blebbing to fool the immune system into uselessly chasing those bits of its former self around while the now cell wall deficient form of the spirochete burrows to safety within the host's own cell walls for protection. And there it will wait- sometimes for years- before it ever reverts to a spirochetal form again. This may happen when the host is re-infected or when the immune system is weakened by things like stress, sickness, nutrient deficiency, or the accumulated affects of aging. [48]

# INTRODUCTION

Humans, and the animals that they have become closely associated with, enter the Lyme disease picture when they become blood meal hosts for ticks. If they interact within a tick endemic area they may become infected, and a small percentage of those who are infected may become symptomatic. The tests used to measure this infection vary. Researchers have squabbled long and hard over their accuracy, but most tests of blood from any particular human population show that a percentage are seropositive with antibodies to a *Bb* infection at all times- with borrelia species specific to specific areas.

Most infections in the United States are with the *Bb* species. In Europe, several other species also cause infection. The life and transmission cycle for other species of *Borrelia* are somewhat different and cause a variety of named ailments, the most important being relapsing fever. All involve an insect or arthropod vector and blood meal host. In Europe *Ixodes ricinus* ticks carry *Borrelia afzelii* and B*orrelia garinii,* which may be more likely to cause neurological symptoms. The *Borrelia miyamotoi* causes illness in Japan. In western Africa, *Borrelia croceroi* is transmitted to humans through the bite of a species of soft-bodied mole tick and causes *tick* or *famine fever*. The prolonged drought of recent years in parts of western Africa has caused a surge in tick borne disease by concentrating mice and ticks in riparian areas where water and water dependant humans are also located. [49]

Drought causes animals to move around in search of water. One study compared the number of mice found along a hiking path during a year of normal rainfall with the number found during a drought year. The number decreased from more than 200 to 2 mice. It is theorized that the mice had either died or, more likely, simply moved to another area with higher moisture levels or access to water-somewhere near a river or lake. In fact riparian corridors, sandy soil, and wooded vegetation have been found to be strongly correlated with clusters of *Ixodes* ticks and Lyme disease foci. During droughts, tick presence on mice and deer is associated with proximity to rivers.[50] The prolonged drought suffered by the Massachusetts Bay Colony during the 1680's and early 1690's may have had a similar concentrating effect on the mice and ticks along the Ipswich River in western Salem Village.[51]

Asymptomatic Lyme disease infections are quite common. It should be emphasized that only a small fraction of those who test seropositive ever have symptoms and that the symptomatic can be seronegative, so this is still a confused and murky picture. Seropositivity can start at birth. One study of obstetrical patients and their infants in endemic areas found that an average of eight percent of all newborns had cord blood that tested positive for Lyme antibodies. None of these infants, however, showed any symptoms at birth. Another study of infants born in Lyme endemic areas of New York State tested at a cord blood level of 10.2 percent seropositivity for antibodies for Lyme disease. [52]

The seropositivity rate for domesticated animals in endemic areas has also been studied. There is considerable variation on a worldwide basis. Cows can become affected and calves can be born seropositive.[53] Pigs, especially wild boar, have been found to harbor voluminous numbers of ticks but no research has been done on their role in the ecology of Lyme infection. In historic times, pigs were ubiquitous creatures of the woods but are generally absent from the oak forests of modern times.[54] Dogs are prone to infection. One study found that seventy-five percent of all dogs in one area of Connecticut were Lyme seropositive, but of these, only five percent were symptomatic. [55]Another study found thirty-six percent of all cats in another town in Connecticut to be seropositive, again with only a few ever showing any symptoms.[56]

Based on these world wide studies of both humans and animals, and considering the lack of reliability of any of the blood tests currently being used, Lyme infection should be looked at as a historic, pandemic, and omnipresent force. Some forms of these bacteria were highly likely, at any point in historic time, to infect a percentage of all the people and animals in any endemic area. A small percentage of them would become symptomatic for a variety of reasons. This would ebb and flow through time in response to changing environmental and cultural practices within various human societies, as well as varying bacterial loads, genetic sensitivities and technical changes like the use of antibiotics and vaccinations.[57]

What would trigger symptoms and who would display them? Using the logic that a portion of any population living in an area endemic for *Ixodes* ticks and *Bb* would be infected at any point in historic time, the triggers to also look for might be stress of some sort: environmental, emotional, social and/or physical stress (see Figure 5). Social conflict, warfare, drought, famine, population movements into new edge environments, edge environment expansion, and emotional distress would also be examples of potential triggers. Tick risk would be affected by masting, moisture levels, the movement of blood meal hosts, and other environmental conditions, as has been implicated in modern studies. It has been proposed that the global warming that we have experienced since the Industrial Revolution may have helped accelerate tick population growth due to a more successful survival

# INTRODUCTION

rate in winters with mild temperatures but this has not yet been supported in studies and has yet to be proven conclusively.[58]

When would people show symptoms? Ticks are most active during certain times of the year. Since infection is not passed from the mother tick to her eggs, the larvae stage is infection free but this does not last. It is the adult and nymphal forms that carry *Bb* infection. They are abundant during specific times of the year. This seasonality has been graphed for modern *Ixodes scapularis* populations in Maine.[59] The tick life cycle would have followed a similar pattern in the historic past with minor revisions for calendar, geographic, and climatic shifts. Most of the symptoms of infection in the historic past would probably have occurred at the times of year immediately following the peak periods (see below) for nymphal and adult tick activity.

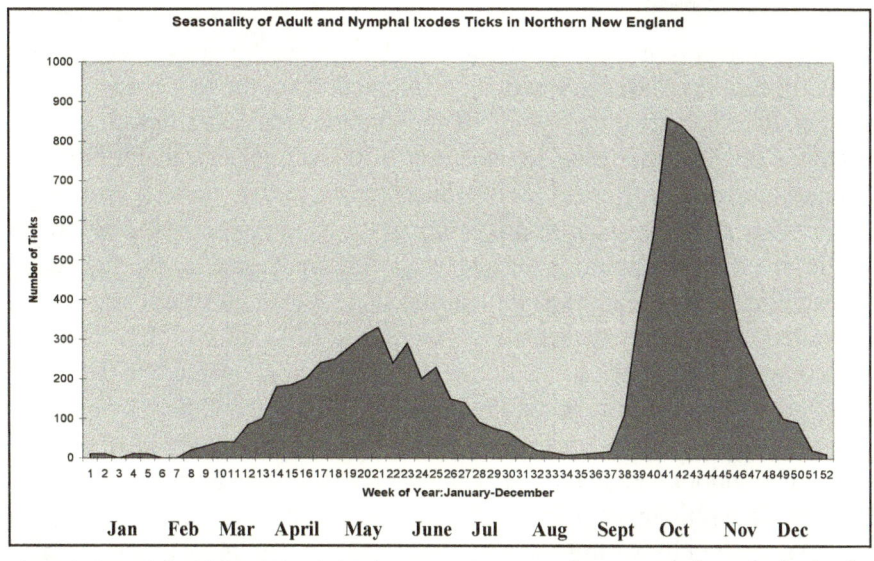

**Figure 3**. *CHART A: Adult and Nymphal Ticks pass on Bb infection.* Average number of submitted ticks in the State of Maine over the past ten years- shows a strong spring and autumn seasonality for the appearance of ticks. Statistics from the MAINE CDC

Who would show symptoms in the past? Modern statistical studies in the United States have found that symptomatic cases of Lyme borreliosis tend to be bimodal-affecting the young, especially those less than 14 years old, more heavily with another peak later in the age span in the older age groups [See Chart B]. This pattern has been duplicated in various European studies. Throughout time and across cultures children may tend to be inherently attracted to play in edge environments that are tick risky and may receive more ticks bites. A young immune system may also be more sensitive to infection, which creates the early incidence peak on the graph. However, humans will only tolerate a certain number of unnoticed tick attachments. The anesthetic like substance that ticks secrete in their saliva to numb the place where they bite only works a few times before a human host's immune system becomes sensitized to it and begins to develop anti-tick saliva antibodies (ATSA). Subsequent tick bites are felt because they become itchy. These antibodies are thought to also cause attached ticks to detach sooner, which may also lower the level of any bacterial transfer. One study has shown that after about three tick bites, the risk of acquiring Lyme disease decreases dramatically. It appears that the tick either attaches for a shorter time or the bite is felt and the tick is simply scratched off. In either case, the bacteria's opportunity to transfer into the human bloodstream from a tick is reduced.[60]

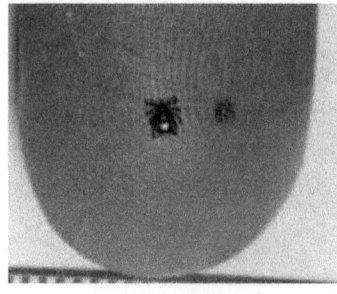

**Figure 4.** *Ixodes ticks.* CDC PHIL

This effect may cause the valley in the graph: after being bitten as a child, the mature human gains a temporary advantage over the tick and the bacteria does not get passed on at the same rate. This effect may also be limited. The anesthetic sensitivity may wear off- perhaps after fifteen years or so-allowing ticks to become stealth infectors once again. Re-infection comes at a time when the human has begun to experience the affects of aging, may be more vulnerable physically and may have adopted an indoor sedentary life. The cyst and cell wall deficient forms of the bacteria within an asymptomatic individual, inactive for years, may also be activated by various forms of stress that accompany aging to create renewed infection. People who dress protectively, wearing long sleeves and pants that were tucked into their socks would show fewer symptoms. People who employed insecticides and repellents would also have fewer symptoms.

A *bull's eye* rash, if it appeared, would show up within a week or two of the tick bite with possible reoccurrences throughout the infection. Arthritic and neurological symptoms usually take a longer time to manifest. Virulence would vary over time and between locations as the *Bb* bacteria evolved and changed.

# INTRODUCTION

## Factors Associated With Risk for Lyme disease and its Associated Manifestations[61]

| Predisposing Factors | Precipitating Factors | Perpetuating Factors |
|---|---|---|
| HLA DR2, HLA DR 4 Genotype | Tick bite (initial infection) | High bacterial load |
| Compromised immune System | Episode of acute stress (relapse) | Re-infection |
| Co-infections that cause Immunosuppression | Immunosuppression (relapse) | Co-infections |
| Ecosystem that fosters Tick-borne disease (infection) | Vaccination (relapse) | Chronic unremitting stress |
| Outdoor activities/ indoor lifestyle (infection) | Childbirth (relapse) | Sleep deprivation |
| Accident (relapse) | Corticosteroid Exposure | |
| | Co-infection (relapse) | Misdiagnosis |
| | PTSD (relapse) | Undertreatment |

**Figure 5.** *Some of these characteristics can be looked for in the past.*

Past human populations would probably have little concern for the *Ixodes* ticks that in their nymphal stage are only a little bigger than the period at the end of this sentence. They would probably have been virtually unaware of their presence in their environment. Attached adult ticks might have been noticed but not recognized as attached ticks. After all, it took modern science several years to figure out the tick-bacteria-affliction connection and that research is still very much a work in progress. The "germ theory of disease" was in its infancy in the

## DISGUISED AS THE DEVIL

seventeenth century. For every Cotton Mather living in New England who was looking in wonder through one of the very first microscopes at a drop of water full of what he described as "eels,"[62] there were thousands who were not. But the afflictions caused by the invisible world of bacteria would have been difficult to ignore.

Figures 6. *Charts of Lyme Disease Cases and Witch Trial Participants*

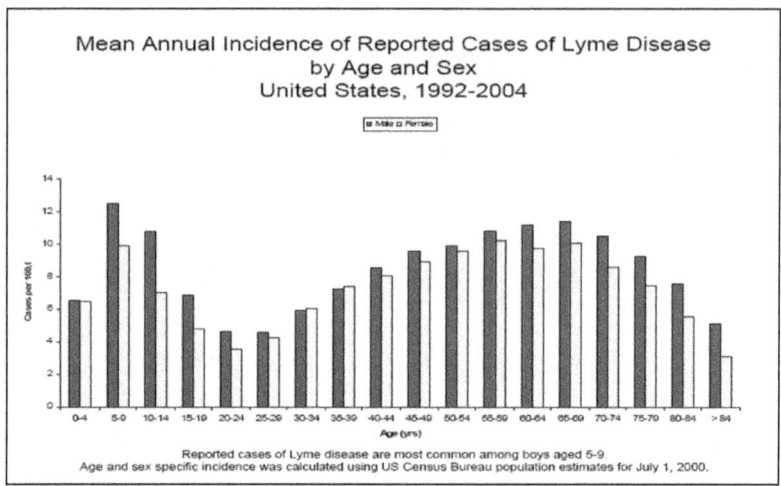

***CHART B.*** *A bi-modal distribution-The young and the older are more likely to be symptomatic.* CDC.

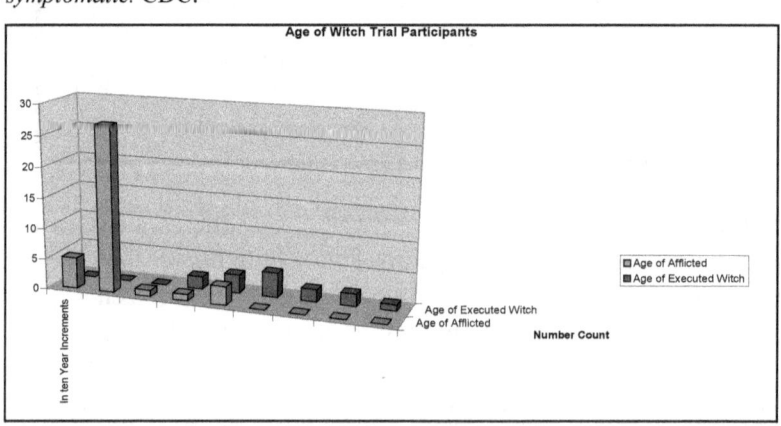

***CHART C.*** *A similar bi-modal distribution-In the Salem Witch Trials of 1692, the young [at left of chart] were more likely to be afflicted, the older –peaking in the 50's age group- were more likely to be executed as witches.* Statistics compiled from Salem Witch Trial Records, Essex County, Vol. I and II, the age of 4 executed witches and 6 of the afflicted are unknown and excluded.

## INTRODUCTION

# INTRODUCTION

[1] The Spanish colonies in the New World continued Spain's history of little interest in witch hunting although in New Mexico, a Spanish settlement, a few witchcraft accusations which appear to be related to Native American shamanism were recorded. It is, however, not a modern *Ixodes scapularis* endemic area. Surprisingly French Canada was also fairly free of witch-hunts, even though it was certainly popular back home in France. This may be because Canada was too cold during the Little Ice Age for an *Ixodes* population to be well established, although that is now changing.

[2] David Grann, "Stalking Dr. Steere Over Lyme disease," *The New York Times Magazine* (June 17, 2001) archives at www.nytimes.com

[3]. Dr. Steere originally included a third stage for Lyme disease in his earliest descriptions of his research. At some point in time his opinion changed and he is now one of the major proponents for a short course treatment of antibiotics for those who contract the disease – if they don't work patients are then classified as having fibromyalgia syndrome, a painful chronic affliction of unknown etiology or post-Lyme syndrome.

[4] A.C. Steere, E.J. Dekonenko, V.P. Berardi, L.N. Kravchuk. "Lyme borreliosis in the Soviet Union," *The Journal of Infectious Disease* 158 No.4 (October 1988), 748-753. This study was done after the Lyme spirochete had been identified and antibiotics were known to be effective in treating the disease. Dr. Steere seems to have been initially resistant to the use of antibiotics for treatment. Nowhere in this sad study does it mention that the patients involved were ever given antibiotic treatment. It is doubtful, but not impossible, that such a study would now be allowed in any western medical system.

[5] This statement was repeated in much of the literature about Lyme disease that was written in the 1990's, it was removed from Center for Disease Control website after the relatives of the deceased complained but is still included in the *Lyme disease* (Atlanta: U.S. Publishing Office, 2001) pamphlet.

[6] See fatalities in K.B. Leigner and C.R. Jones, *Fatal progressive Encephalitis*, Abstract presented at the International Conference on Lyme disease and other Tick Borne Diseases at Munich, Germany (June 1999).

[7] CDC, *Lyme disease*.

[8] *Public Law 107-116*, was signed by President Bush on 1/10/2002.

[9] Joel M. Shmukler, *Safety/Efficacy Concerns Re: Lyme Vaccine: Lymerix* at www.lymeinfo.net.

[10] This issue is outlined in the Lyme Disease Association's Position Paper *Conflicts of Interest in Lyme disease: Laboratory Testing, Vaccination, and Treatment Guidelines* (April 2001) available at www.lymenet.org.

[11] *Ibid*, Section XII, subsection 5.

[12] A.C. Steere, E. Dwyer, and R. Winchester, "Association of Chronic Lyme arthritis with HLA-DR4 and HLA-DR2 alleles," *The New England Journal of Medicine* 323 (July 26, 1990), 219-223.

[13] Isabel Diterich, Carolin Rauter, Carsten J. Kirschning, and Thomas Hartung· "*Borrelia burgdorferi*-Induced Tolerance as a Model of Persistence via Immunosuppression" *Infect Immu.* 2003 July; 71(7): 3979–3987.Yeremeev VV, Lyadova IV, Nikonenko BV, Apt AS, Abou-Zeid C, Inwald J, Young DB. "The 19-kD antigen and protective immunity in a murine model of tuberculosis" *Clin Exp Immunol.*.May 2000;120(2): 274-9.

[14] Lyme Disease Association, Section IX, "Size of the Lyme Disease Market in U.S. Dollars."

[15] See National Institute of Allergy and Infectious Disease *News Release* (12/04/2004) "Broad- based Vaccination of Wild Mice Could Help Reduce Lyme Disease Risk in Humans" at www.niaid.nih.gov.

[16] F.R. Matuschka, Endepols, S., Richter, D., Ohlenbusch, A., Eiffert, H., Spielman, A., "Risk of urban Lyme disease enhanced by the presence of rats," *J Infect Dis.* 174 (1996), 1108-11.

[17] See A. Eisenberg, *News brief: Biology professor sheds light on Lyme disease*, 12/04/2007 at www.dailypennsylvanian.com

[18] Margherita T. Cantorna, Colleen E. Hayes, and Hector F. DeLuca, "1,25-Dihydroxycholalciferol Inhibits the Progression of Arthritis in Murine Models of Human Arthritis," *The Journal of Nutrition* Vol. 128 No. 1 (January 1998), 68-72, and Ronald Schell, *Osp A Induces Lyme Arthritis in Hamsters*, presentation at the 12th Annual Lyme disease Foundation Scientific Conference (April 9, 1999).

[19] Lyme Disease Association, op. cit., Section IX, "Size of the Lyme Disease Market in U.S. Dollars."

[20] CDC, *Lyme Incidence Report* for 2002.

[21] The two standards of care and the ILIADS position and differences with IDSA are included in "International Lyme and Associated Diseases Society (ILADS) criticizes just-published article in the New England Journal of Medicine: Article concludes Chronic Lyme Disease a misnomer, despite significant real-world evidence proving otherwise" *Press Release*, Bethesda, MD (PRWEB) October 3, 2007.

[22] Gregory A. Poland, "Prevention of Lyme disease: A review of the Evidence," *Mayo Clinic Proc.*76 (2001), 717-19.

# DISGUISED AS THE DEVIL

[23] See Vos, K.,Van Dam, A.P., Kuiper, H., Bruins, H., Spanjaard, L., Dankert, J., "Seroconversion for Ld borreliosis among Dutch Military," *Scandinavian Journal Infectious Disease* (1994).

[24] EPA R.E.D. Facts DEET EPA-738=F-95-010 April 1998. See also a Nazi project described in John Loftus' *The Belarus Secret: The Nazi Connection in America* (New York: Knopf, 1982) that studied dropping infected ticks from airplanes as an early form of biological warfare, and a later Soviet tick "weaponizing" program.

[25] See, for example, www.actionlyme.org, Kathleen Dickson's case may be an extreme one but it highlights the challenges involved with Lyme disease. Kathleen has testified as an expert witness before the F.D.A. She proclaims a long list of problems attaining treatment and long list of accusations against the Connecticut Medical establishment. She has filed a racketeering complaint against several drug companies and physicians for conspiring to change the definition of Lyme disease.

[26] See Robert Calef, " New Wonders of the Invisible World," in George Lincoln Burr's *Narratives of the New England Witchcraft Cases* (Mineola, New York: Dover, 2002).

[27] Betty Parris grew up, married, and lived until 1760. Her brother Noyes, however, is reported to have died young and was described as suffering from mental illness. Abigail Williams is thought to have also died young, perhaps by suicide, see Frances Hill, *Hunting for Witches* (Beverly, Mass: Commonwealth Editions, 2002), 71.

[28] She "gathered flesh exceedinglye.."in Samuel Willard, "A briefe account of a strange & unusuall Providence of God befallen to Elizabeth Knap of Groton" in Samuel A. Green, ed. *Groton In The Witchcraft Times* (Groton, MA: [s.n.] 1883).

[29] The case of the Godwin children is described in great detail in Cotton Mather's "Memorable Providences Relating to Witchcrafts and possessions" written in 1689, Burr's *Narratives*, 101-119,quote from Lawrence Mayo, Ed., Thomas Hutchinson, *History of the Province of Massachusetts Bay,* Vol.2 (Cambridge, Mass, 1936),16.

[30] Ann Putnam made her speech in 1706.

[31] Chamberlain family oral history. Rebecca lived in Billerica, Mass. and died in prison in 1692, having been accused of practicing witchcraft by her husband. At least one descendant has stated that by oral history she had red hair.

[32] Cat ownership has been found to be associated with Lyme disease in humans in two case reports (Curran and Fish, published letter, 1989). See Judy Folkenburg "Pet Ownership: Risky Business?" *FDA Consumer Magazine* at www.fda.gov.

[33] Michael Christopher Carroll, *Lab 257, The Disturbing Story of the Government's Secret Plum Island Laboratory* (New York: Morrow, 2004) for an overview of research at Plum Island. Other theories for the origins of Lyme disease in North America include the Vanderbilt family bringing it into New York Harbor on animal pelts from Siberia, other evidence suggests that the appearance of the disease in predates all these activities.

[34] W. Harvey and P. Salvato, "Lyme disease: Ancient Engine of an unrecognized *Borreliosis* pandemic," *Medical Hypothesis* (2003), 742-759.

[35] Robert Aronowitz, " Lyme disease-the Social Construction of a New Disease and Its Social Consequences" in *Making Sense of Illness, Science, Society and Disease* (Cambridge U.K.: Cambridge University Press,1999), 57-83.

[36] Dr. Kenneth Leigner, "Lyme disease presentation" at the *12th Annual International Conference on Lyme disease and other Spirochetal and Tick Borne Disorders,* New York, April 9,1999, see also Margos G et al. "MLST of housekeeping genes captures geographic population structure and suggests a European origin of Borrelia burgdorferi." *Proc. Nat.l Acad. Sci.* 2008; 105(25): 8730-8735.

[37] John Gerrard wrote one of the earliest English herbal guides, published in 1597, but the most well known of the seventeenth century was probably Thomas Johnson's *Herball - General Historie of Plants*( London: Probably Self Published,1633) which included many cures and protective uses. Modern studies of the herbs and spices used in cooking have found that many are highly antimicrobial, see Paul Sherman and Jennifer Billing, "Darwinian Gastronomy: Why we use spices," *Bioscience* Vol 49 No 6, 456.

[38] See Mather, *Illustrious Providences*, 167. The surname of this family occurs in the records as both Hortado and Fortado. It will be consistently Hortado in this book. Mary Hortado died in a 1690 raid during King Williams War.

[39] Jane Fearnley-Whittingstall, *Ivies*. (London: Chatto & Windus Limited, 1992).

[40] The symptoms of neurasthenia are listed as psychological plus fatigue, weakness, headache, sweating, polyuria, tinnitus and vertigo, photophobia, fear, easy exhaustion on the slightest effort, inability to concentrate, irritability and complaint of poor memory, poor sleep, numerous constantly varying aches and pains, vaso-motor disturbances. From *Taber's Cyclopedic Medical Dictionary* (Philadelphia: F. A. Davis Co, 1952). This could be a classic description of a chronic Lyme patient.

[41] Quote from Freud in S. E. Kris, Introduction; M. Bonaparte, A. Freud, & E. Kris, Eds.; E. Mosbacher & J.

# INTRODUCTION

Strachey, Trans. *The Origins of Psycho-analysis, Letters to Wilhelm Fliess, Drafts and Notes: 1887-1902.* (New York: Basic Books. 1952).

42 See Yvonne Anraham, "The Hidden Victims Of Lyme Disease," *The Boston Globe*, in Lyme Times (1999) at www.lymedisease.org

43 Polly Murray, *The Widening Circle* (New York: St. Martin's Press,1996), 31-45.

44 Geoffrey Cowley and Anne Underwood, "A Disease in Disguise: Lyme can masquerade as migraine, or as madness," *Newsweek* (August 24, 2004), 62.

45 Birds play a key role in the maintenance and spread of Lyme disease. See J. Terborgh, "The role of ecotones in the distribution of Andean birds," *Ecology* 66 (1985), 1237-1246, also see maps of migratory routes, Two other important studies are K. Kurtenbach, et al., "The Key roles of selection and migration in the ecology of Lyme borreliosis," *International Journal Medical Microbiology* 291 Sup.33 (June, 2002),152-4 and J. D. Scott, "Birds Disperse Ixodid and Borrelia burgdorferi infected ticks in Canada," *Journal Med. Entomology* 38 No.4 (July, 2001), 493-500. The stress of migration may activate Bb in birds. See S. Millus, "Migration may reawaken Lyme disease," *Science News* (February 19, 2000).

46 Ian D. Campbell and John H. McAndrews, "Forest disequilibrium caused by rapid Little Ice Age cooling," *Nature* 366, December 1993, 336 – 338 discusses the changing composition of forests in relation to climate change. Beech trees were replaced by oak and then pine trees in southern Canada in response to the cooling of the climate during the Little Ice Age. Oak trees are harbingers of Ixodes ticks.

47 S.E.Malawista and Chevance de Boisfleury, "Clocking the Lyme Spirochete" *PLoS ONE, 3 (*2) February 20, 2008, e 1633.

48 Embers M.E. "Survival strategies of Borrelia burgdorferi, the etiologic agent of Lyme disease." *Microbes and Infection, 6* (2004), 312-318.

49 IRIN News Reuter Foundation, *Senegal: Lyme disease: the forgotten scourge of West Africa*, August 23, 2006

50 Jones, C.J. and U.D. Kitron, "Population of Ixodes scapularis are modulated by drought at a Lyme disease focus in Illinois," *Journal Med Entomol*, 37 (2000), 408-415.

51 See the Witch Drought Appendix D.

52 See Daniel Elliot, S. Eppes and J. Klein "Teratogen Update: Lyme disease," *Teratology* 64(2001), 276-281.

53 N.Popvic, B.Djuricic, and M. Valcic "The Importance of Lyme borreliosis in Veterinarian Medicine," (Translated from Serbian) *Glas Srp Akad Nauka* 43 (November 1993), 277-85. Karen Vanderhoof-Forschner also discusses this topic at length.

54 Greiner, Ellis, Pamela Humphrey, Robert Belden, William Frankenberger, David Austin, and E. Paul Gibbs. "Ixodid Ticks on Feral Swine in Florida,"*Journal of Wildlife Diseases*, 20 (2) (1984).

55 Bogumila Skotarczak, Beata Wodecka, Anna Rymaszewska, Marek Sawczuk, Agnieszka Maciejewska, Malgorzata Adamska, Teresa Hermanowska-Szpakowicz, Renata Swierzbinska, "Prevalence of DNA and antibodies to borrelia burgdorferi senso lato in dogs suspected of borreliosis," *Ann Agric Environ Med* 12 (2005), 199-205.

56 M. Gibson, C. Young, M. Omran, J. Edwards, K. Palma, L. Russell, J. Rawlings, "Borrelia burgdorferi infection of cats," *J Am Vet Med Assoc.* 202 (1993),1786.

57 Robert C. Bransfield, "Lyme Disease, Co-morbid Tick-Borne Diseases, and Neuro-psychiatric Disorders" in *Psychiatric Times* Vol. 24 No. 14. (December 01, 2007).

58 Theoretically the warmer winters caused by global warming will allow more ticks to survive through the winter. Study results have not yet consistently supported this hypothesis.

59 CHART B is based on statistics from the State of Maine for adult and nymphal tick submissions between 1989-2002 to The Maine Vector Borne Disease Lab in Portland, Maine.

60 G. Burke, et al., "Hypersensitivity to Ticks and Lyme disease Risk in Heavily Tick Infested Areas," *Journal of Emerging Infectious Diseases* 11 No.1 (January 2005), 36-41. Being bitten by ticks repeatedly elicits a response that limits the anesthetic feature in tick spit making the bite more likely to be felt and the tick to be scratched off before it can pass on the Bb bacteria.

61 From Robert C. Bransfield, MD, "Lyme Disease, Co-morbid Tick-Borne Diseases, and Neuro-psychiatric Disorders," *Psychiatric Times* Vol. 24 No. 14, December 1, 2007.

62 Cotton Mather's interest in science is discussed in Kenneth Silverman, *The Life and Times of Cotton Mather* (New York, NY: Columbia University Press, 1971).

DISGUISED AS THE DEVIL

# CHARACTERISTICS PRESENT AND PAST

| Modern Lyme Disease Characteristics: Present | Witchcraft Characteristics: +/- 1350 to 1782* |
|---|---|
| May be related to Weather: Temperature/Moisture | May be related to weather: Little Ice Age/rain/drought |
| Population obese, Vitamin D deficient, malnourished due to modern western diet | Population often malnourished due to drought, and crop failures |
| Ticks attach to body to suck blood, Tick saliva numbs the skin prior to attachment<br>Infection transferred when tick attaches and sucks a blood meal<br>$Bb$ is invisible to naked eye. | Pins are stuck in skin, and Devil's mark/witches teat does not have any feeling<br>Witch has familiar that sucks blood from witch's teat<br>Effluvia is an invisible substance used by witches to cause affliction |
| ECM/Bull's Eye Rash/Lymphocytoma in Europe | Devil's Marks/Witch marks/Preternatural Marks on skin (enter literature after 1487) |
| Rural/SUBURBAN/Not Urban | Rural/VILLAGE/Not Urban |
| High prevalence in riparian/watershed zone/ forest edge environment/ glacial outwash/sandy-fertile soils | High prevalence in riparian/ watershed zone/ forest edge environment/ glacial outwash/sandy- fertile soils |
| Oak/Mixed Forest Edge Concentration above 40 degrees Latitude | Oak/Mixed Forest Edge Concentration above 40 degrees Latitude |
| Multi-system symptoms in humans: Neurological with arthritic | Multi-system symptoms in humans: Neurological with arthritic |
| Walking may be difficult due to arthritic pain, May use cane or walking stick | Witches have no need to walk-they fly using a pole, broom or animal |
| Affects humans and Domesticated animals | Affects humans/Crops/ Domesticated animals |
| DEET repels ticks | Herbs with insecticidal qualities deter witches |
| Heaviest affliction in children and older adults | Heavy affliction in young/witch is old hag |

\* *This is the date of last legal execution for witchcraft in Europe.*

Figure 7.

AFFLICTION

# I. AFFLICTION

"...this small, vile creature may, in the future, cause the inhabitants of this land great damage"[63]
Peter Kalm, writing about ticks in 1749.

As has been noted, in North America Lyme disease is usually spread by the bite of ticks of the genus *Ixodes* that in the United States are infected with the *Borrelia burgdorferi sensu lato* (*Bb*) bacteria. These ticks may also be co-infected with other germs. *Anaplasma (Erhlichia), Babesiosis, Bartonella, Mycoplasma* and *Rickettsiae* are all co-infections that are carried by the same ticks. For Lyme disease to be endemic in an area, at least three closely interrelated elements must be present: the bacteria, the ticks that can transmit bacteria, and mammals (such as mice, deer and humans) or birds that provide blood meals for the ticks. Blood is a tick's only form of nourishment.

*Ixodes* ticks are usually found in temperate regions with high relative humidity at ground level. In New England, *Ixodes scapularis* ticks are associated with deciduous forest and habitat containing leaf litter. Leaf litter provides a moist cover that protects from wind, snow, and other elements. *Ixodes scapularis* ticks are generally found in the edges between heavily wooded areas and tracts of cleared land--the transitional ecotone that is also favored by the white tailed deer. In areas where most of the land is cleared, ticks may be limited to riparian areas that are regularly visited by thirsty host mammals.

Knowing the complex life cycle of *Ixodes* ticks is important in understanding the risk for humans acquiring Lyme disease. In New England *Ixodes scapularis* has a two-year life cycle and three life stages. Adult female ticks lay eggs on the ground in early spring. By summer, the eggs hatch into larvae. Larvae feed by attaching to mice, other small mammals, and birds in the late summer and early fall and then molt into nymphs and are inactive until the next spring. Nymphs feed on rodents, small mammals, birds, and humans in the late spring and summer and then molt into adults in the fall. In the fall and spring, adults feed and mate on large mammals (usually deer) and can bite humans.

## DISGUISED AS THE DEVIL

The key roles that deer play in this cycle is that they can nourish a large number of ticks, assuring population viability and stability, and, the ticks are moved around on their backs from place to place. Where the deer are is geographically also where the ticks will be most heavily concentrated. However, deer do not spread the Lyme bacteria. Deer have an antibody in their blood to the outer surface proteins of the *Bb* spirochete (their health is usually not affected) which may act somewhat prophylactically to eliminate the *Bb* bacteria in the adult ticks during their last blood meal. The adult females that drop off lay bacteria-free eggs in the spring, which completes the two-year life cycle. In the majority of cases, larvae hatch in a bacteria free state and are infected only after contact with an infected blood meal host.

Ticks, birds, rodents, and some other animals all serve as natural reservoirs for the *Bb* bacteria. This spirochete can usually live and grow within these hosts without causing them to die. Tick larvae and nymphs generally become infected with the *Bb* spirochete when they feed on small animals and birds that carry the bacteria. Once infected, the bacteria remain in the tick for the rest of its life until the final deer blood meal. Infected ticks bite and transmit *Bb* to rodents, birds and other mammals in the course of their normal feeding behavior.

Ticks search (or quest) for host animals from the tips of grasses, shrubs, and leaf litter and transfer to animals or people that brush against this vegetation. Ticks can only crawl. They do not fly. Ticks can attach to any part of the body but will often crawl to a moist hidden area like the groin, armpit, or a place near the waist where the skin may be sweaty. They feed on blood by first applying a sort of spit anesthetic to the host's skin to numb it and then attach their mouths into the skin itself. Their bodies slowly enlarge as they become engorged with blood. This can take several days.[64] This complex process, life cycle and seasonality have been occurring in New England for a long time. When early human settlers landed and interacted with their new habitat they became just another set of unwitting blood meal hosts for a pre-existing *Ixodes scapularis* tick population.

Like so many other features of New England's landscape, the prevalence of Lyme disease is directly influenced by both nature and the 'hand of man.' Glaciation related population fluctuations affected both the *Bb* bacteria and the *Ixodes scapularis* tick vector on a genetic basis. High prevalence rates for Lyme disease skirt northern latitudes on a worldwide basis in areas with deciduous mixed forests that usually have a high percentage of oak trees. These are areas that were influenced by the

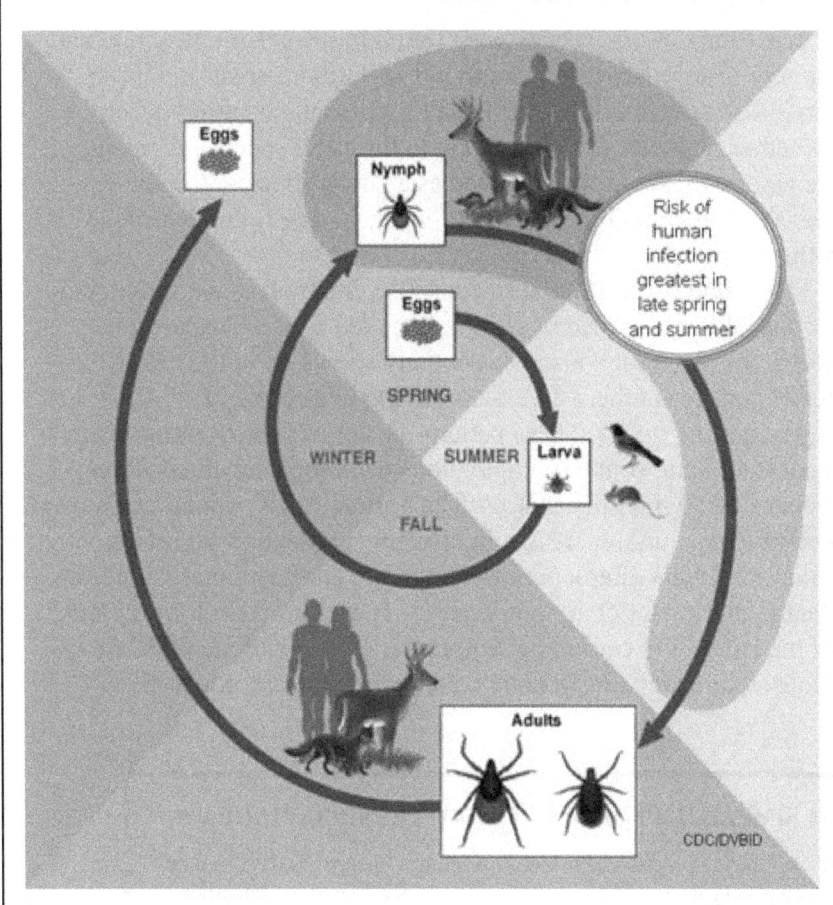

**Figure 8**. *Ticks have three chances in their life cycle to acquire Borrelia burgdorferi bacteria.* CDC

last glaciers that occurred between 18,000 and 15,000 years ago. It has long been observed that the species diversity of many animals and plants decreases with higher latitude. Glaciations may have played an important role in this phenomenon.[65] Because of its importance as a pathogen of humans and the value of complete genome sequence information for understanding its life cycle, the *Bb* bacteria were subjected to a gene sequencing study in the late 1990's. It contains a linear chromosome of 910,725 base pairs.[66] Studies have found a relative genetic homogeneity among *Bb* bacteria species in northern sections of North America which is indicative of an evolutionary history of recent (15,000 years or so) population growth and rapid geographic range expansion for this species.[67]

## 🐢 DISGUISED AS THE DEVIL

In the genetic study of the *Ixodes scapularis* tick in North America, substantial differences were found in the genetic structure and evolutionary history between ticks found in northern areas and those found in inland southern areas. This underscores the ongoing debate about whether both sets of ticks belong to the same species. Although they are taxonomically now described as a single species, *Ixodes scapularis*, two distinct mitochondrial lineages have been identified within the species.

One lineage, the "Southern clade," is common in the south, and another, the "American clade," (formerly called *Ixodes dammini)* is common in the north (and also occasionally shows up in the mixed genetics of coastal southern area ticks having been brought down by migrating birds). These patterns of genetic composition suggest that the northern and southern populations have separate and different evolutionary demographic histories. The northern tick population showed evidence of exponential increases in effective population size. This population explosion after a period of extreme contraction is sometimes described as an 'evolutionary bottleneck.' The contraction forming the neck of the bottle was caused by temperature stress and die off during glaciations. Corresponding genetic patterns of southern population

---

**Figure 9. SIMPLIFIED POPULATION BOTTLENECK DIAGRAM**

Northern lineage Ixodes scapularis    Southern lineage Ixodes scapularis
If D is a good carrier of Bb                    (Not affected by glaciation)
                         TIME

| more prevalence | | less prevalence |
|---|---|---|
| C D CC D C D D C | present | A C D GG FBGH C A |
| DC DCC DCCD C | | BCDEFGHHDCBAHC |
| CDCDCDCD | | HCACHGBBFGDCAH |
| DCDC | | ADCGGFBHCADAHC |
| CD <~~~~~ | glaciation | CDHGBFGDAAHCD |
| CDHC | | DEFGHABCDEEGHC |
| DHCGFEC | | EGDAHDBBCCDDHC |
| ABCDEFGHAD | | ABCDEFGHABCDEF |
| ABCDEFGHABCDEF | | ABCDEFGHABCDEFC |
| | past | |

samples showed a long evolutionary history of constant and stable population size.[68] The localization of Lyme disease in the range of the Northern lineage of *Ixodes scapularis* suggests that this genetic factor may play an important role in the concentration of this disease in the north. The lower rate of *Bb* infection in southern *Ixodes scapularis* could be the result of the more diverse genetic heterogeneity of southern ticks as well as environmental conditions.

A study of a forest/tundra ecotone in modern Colorado found that tree establishment is correlated with climate change. In the past, the warming of the atmosphere was accompanied by the melting of glacial ice and environmental change.[69] The glaciations of North America created a slowly moving set of temperature related transitional zones, or ecotones, over time. At 18,000 years before the present (BP), Massachusetts was tundra at the edge of a glacier. This, in turn, edged a larger swath of conifer-dominated zones that covered most of the eastern seaboard, the Midwest, and the upper southern area. The humid temperate/conifer mixed woodland that is good *Ixodes* tick habitat was then restricted to two zones on what is now the southern Gulf Coast. Open temperate woodland also developed in the west.[70] As has been suggested for modern climate changes, as the atmosphere warms up prevailing ecotones move north.

The genetic consequence of tick population contraction during the glacial maxima and population expansion during the intervening warming periods for northern ticks created a discernable morphological change and loss of genetic variation, especially when compared to the diversity of southern populations that were not affected. One recent study has proposed the idea that ecotones are an important source of speciation, which appears to have almost happened to northern tick populations [For many years, the northern clade was called *Ixodes dammini* and the debate over whether it is a truly separate species has been a lively one].[71] Pleistocene glaciation events seem to have caused parallel biogeographic patterns of reduced genetic variability in the north for a wide range of animal and plant species on a worldwide basis. Various genetic studies of northern animal populations have often found the same short-spanning star phylogenies among haplotypes (differing from each other with one or two nucleotides) that is observed in northern *Ixodes scapularis* ticks. [72]

Biodiversity plays a key role in the spread of Lyme disease. When ticks have a variety of blood meal hosts they maintain a lower rate of *Bb* infection, due to the variability of host transmission capacities for the bacteria. This is called a Dilution Effect Model. Lizards (especially in the

 DISGUISED AS THE DEVIL

south), opossums, and squirrels, for example, are poor *Bb* transmitters. But, in reverse, when ticks feed on a limited number of species of animals that are good transmitters (mice and birds) there is a dramatic rise in the level of *Bb* infected ticks.[73]

The 'hand of man' has helped to decrease the number of species living in New England today. Certain species, like the passenger pigeon, were hunted to extinction. Other species, like the white tailed deer and the beaver, tottered close to decimation and then rebounded. The domesticated pig played a role in the ecology of the seventeenth-century forests by competing with indigenous animals for food and territorial space and rooting the forest floor clear of shrubby undergrowth.

**Figure 10.** *Steep stony hillsides like this one were too difficult to plow. They remained wooded and potential tick habitat, although animals were often pastured amongst the trees. This slope, found in modern Danvers, Massachusetts, has never has anything built on it.* Collection of the author.

Forested landscapes have been carved up and fragmented and biodiversity has decreased.[74] New England's forest was extensively altered by Native Americans, eventually deforested by European settlers, but has become reforested during the past 150 years due to farm abandonment, often in a fragmented pattern. It is estimated that the forested areas of New England are now at a level of acreage equal to that extant during the Revolutionary War period.[75]

Two modern studies have found a relationship between forest fragmentation and disease. Research done at the University of Florida found that fragmented forests had higher concentrations of parasites and that animals living within the fragments suffered from a higher level of infection than those living in undisturbed areas.[76] This may be related to an almost unavoidable interaction with the stressors of an encircling edge environment. A study of birds in western Minnesota, for example, found that brood parasitism was higher in nests located near a wooded edge than those located far from an edge.[77] Another study found that the density and infection rates for *Ixodes scapularis* ticks with Lyme disease was dramatically higher in small forest fragments. Fragments that totaled less than five acres carried a Lyme disease risk that was seven times greater than that found in larger areas. This was also found to correlate with a high population of white-footed mice in these same forest fragments.[78]

Like everything else involved with Lyme disease, explanations for the episodic and epidemic level of virulence in New England among humans, ticks and mammals is complex and still subject to much study. But there seems to be a direct link with both biotic and genetic deprivation caused by glaciations, human activities like hunting, deforestation, and reforestation and weather and climatic conditions. In response, New England's *Bb* bacteria have evolved to specialize on populations of genetically similar hosts, which increases virulence and creates an elevated disease incidence rate. Against this background, if those first English settlers visited modern Cape Cod, the wooded areas outside Boston, or suburban Connecticut today, and did exactly The same things they did back then, wearing exactly the same type of clothing that was worn more than three hundred years ago, there is a high likelihood that they would have come into contact with *Ixodes scapularis* ticks and the *Bb* bacteria that they carry. Although the landscape has undergone major transformations over time, there are areas that can be identified as being most likely to have been inhabited by ticks in the 1600's. These areas occurred in marginal zones between the woods and shore, in the mosaic of field and wood, under oak trees, and along paths used by deer. Later, when those "first" settlers chose a site that was "fit for situation"[79] at Plymouth or chose the most defensible sites around Boston Harbor, they chose sites that were undergoing the natural process of re-forestation.

These areas had been previously cleared of trees by Native Americans inhabitants who died in waves of epidemic sickness after contact with Europeans. The "Columbian Exchange" of diseases that transferred germs from Europe to North America was particularly deadly was particularly deadly in New England. It is estimated that the Native population of the Massachusetts had gone from a pre-contact high of +/-

DISGUISED AS THE DEVIL

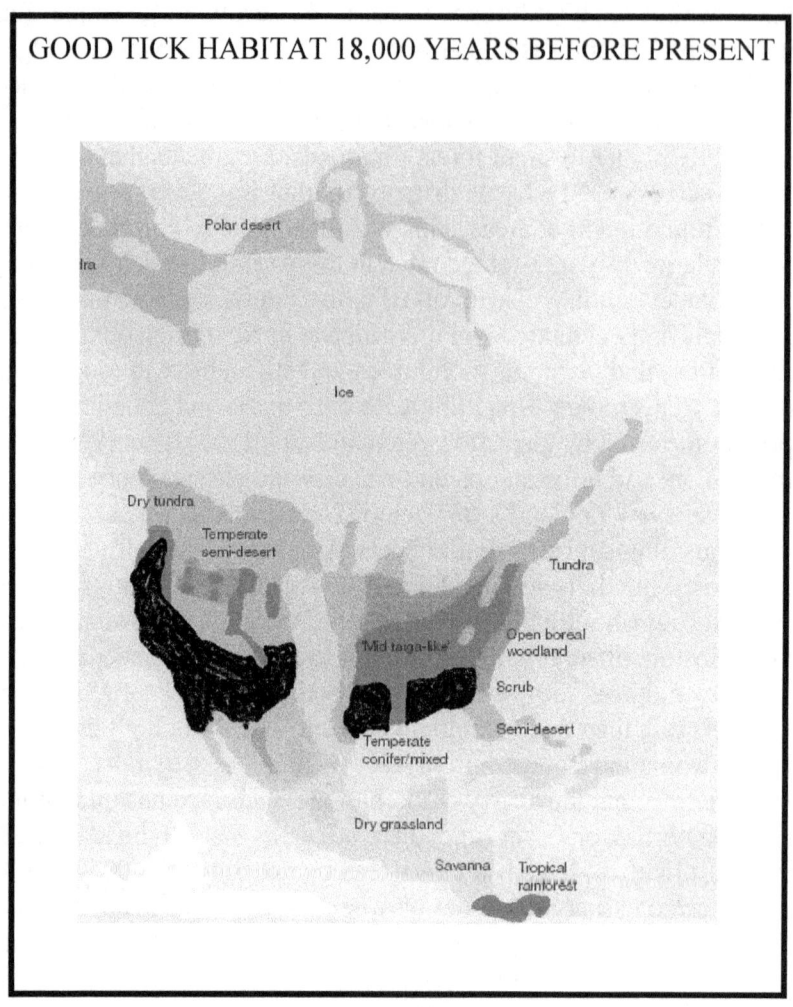

**MAP 2**: *North America during Glaciation at 18,000 years before present [Modern shorelines shown] Massachusetts is tundra at the edge of ice at this point. Environmental ecotones move up and down with the advance and retreat of glaciers. Note that the temperate mixed forest that is good habitat for Ixodes ticks is limited to two mixed zones depicted in black on either side of the proto- Mississippi River and the temperate open woodland in the west.*

# AFFLICTION

9,000 to only about 900 by 1620.[80] The indigenous population would never rebound. The entire village on the land that would become Plymouth Plantation died of disease in a 1616-17 plague.[81]

This meant that at least three years worth of brushy vegetation was growing on that site. Edward Winslow described weeds that "were as tall as a man." The immediate adjacent area would also not have been subjected to at least three years of "burning," the Native practice for clearing brush and fallen leaves in the vicinity of their villages. Modern studies show that repeated burning is an extremely effective tick control measure.[82] However, it also creates an area of luscious new growth in the following spring that is highly attractive to deer. The deer population in the area was described by early observers as being "abound," probably because of the lack of Native hunters and the continued presence of both mast (acorns), the deer's preferred food, and good habitat conditions.[83] Re-forestation creates a marginal brushy edge zone that, when coupled with a supply of birds, mice, deer and ticks, is part of the modern "geography of risk" for Lyme disease.[84]

Modern scientists use several factors to predict the amount of risky 'tick friendly' habitat a geographic area contains. In Wisconsin, tick densities were highest in moist but well drained areas with sandy soils over underlying sedimentary bedrock. Because the immature stages of *Ixodes scapularis* over winter in topsoil and leaf litter, they prefer a forest like that in early Massachusetts that includes many deciduous trees that drop their leaves in the fall. The brushy transition zone under trees and between forest and cleared land is most likely to harbor questing ticks. In areas where land is cleared and especially during droughts, tick populations are concentrated in river corridors.[85]

The following recommendations for the prevention of Lyme disease made by the Massachusetts Department of Public Health in 2001 would have been difficult or almost impossible to follow in *the seventeenth century*:

1. Educate your family about tick-bite prevention. *In the seventeenth century, even if the colonists saw these teensy ticks or found one attached to their skin they did not yet connect insects with disease transmission,* 2. Dress properly--wear light colored clothing, long sleeves, tuck everything in. Wear a hat. Wear tightly woven clothing. Wear shoes with covered toes. Clean your clothes immediately and properly using hot

## DISGUISED AS THE DEVIL

water and a hot dryer after potential contact. *Seventeenth-century men conformed somewhat with tucked in layers; women and children wore long untuckable skirts. Clothing was neither immediately nor properly cleaned,* 3. Avoid contact with tick-infested vegetation. *During the exploratory initial phase of settlement the opposite was done.* 4. Conduct tick checks anytime you have been in a tick habitat. *Non-engorged deer ticks can be the size of the period at the end of this sentence and are hard to see even if anyone was actually looking for them*, and, 5. Use tick repellants and pesticides. *Some herbals may have been used as insect repellants but DEET was three centuries away from being available.*[86]
Lyme disease is caused by the transfer of the *Borrelia* spirochete from an infected tick to the blood supply of a host bird or animal. In any endemic area it is possible that a percentage of the human population is infected by the spirochetes at all times but symptomatic infection may require a full set of co-existing factors to co-occur. Only a portion of those who are infected ever become symptomatic. One early study by Dr. Alan Steere found that for every person in an endemic area who exhibits symptoms, there is another person who has been infected but displays no symptoms whatsoever.[87]
This anomaly may be related to the growing modern problem of vitamin D deficiencies. Vitamin D is created in the human body when sunlight interacts with the human skin. It can also come from dietary sources. Although milk has been supplemented with vitamin D for a long time, rickets, the disease of D deficiency, has begun to reappear in America in recent years.[88] Lyme disease may be associated with this sunlight/skin interaction because its prevalence is concentrated in the northern latitudes, especially above the forty-degree latitude line where sunlight levels are low during the winter.
 The modern "appearance" of Lyme disease coincides with the twentieth century's public health "sun safe" campaigns and "heart healthy" campaigns which advised the avoidance of sunlight and saturated fat. Lard became a taboo substance. Historically, lard was the primary dietary source for vitamin D in many traditional European diets.[89] When rendered from pigs that have access to sunshine, lard contains a whopping 2800 units of D per ounce and is second only to cod liver oil, a traditional winter tonic, as a rich source of vitamin D.[90] Americans have been told to eat less fish, another good source of vitamin D, due to mercury pollution.[91] When put all together these recommendations have had the side effect of creating a vitamin D deficient American.

## AFFLICTION

Lyme disease has a seasonal pattern of infection that has already been linked to the life cycle of the *Ixodes* tick, but may also be related to seasonal blood serum levels of vitamin D. In the Northeastern United States, the highest Lyme disease infection rates begin in the spring when humanity has their lowest post-winter vitamin D blood serum levels. This spring and early summer prevalence occurs even though there are sometimes many more potentially infected adult and nymphal ticks around during the late summer and fall. This may be because human blood serum vitamin D levels with their immune stimulating properties are then at their peak in the autumn.

Research has shown that vitamin D plays a role in the ability of the immune system to provide a rapid response to repel assaults from numerous infectious agents, including bacteria. Vitamin D has been found to induce the expression of the human cathelicidin antimicrobial peptide gene in human bone marrow cells. Vitamin D and Vitamin D Receptors (VDR) play an important role in the regulation of immunity.[92] Another study found that the mechanism for this role in innate immune responses is the activation of Toll-Like Receptors (TLR) that trigger direct antimicrobial activity against bacteria. TLR activation in a human model up-regulated expression of the VDR and the vitamin D-1-hydroxylase genes, leading to induction of the antimicrobial peptide cathelicidin, which (in this study) killed the bacterium tuberculosis. Differences in human population vitamin D levels may contribute to varying levels of susceptibility to all bacterial infections, including Lyme disease.[93]

In another study, vitamin D has been found to decrease the severity of Lyme arthritis and even halt the progression of severe Lyme arthritis in laboratory mice. In the sunlight-deprived laboratory setting, two groups of mice were injected intra-dermally with *Bb* spirochetes and they all developed acute arthritic symptoms within seven days. One group of mice was fed a diet that contained vitamin D. The control group was fed a diet that did not include any vitamin D. The group that was fed vitamin D had dramatically decreased severity of arthritic symptoms when compared with those on a D-less diet.

In the same study, a second set of mice were also infected with *Bb* to induce arthritis. In this part of the study, the mice were treated individually as they developed arthritic symptoms, with one group receiving an inter-peritoneal injection of vitamin D and the control group receiving an injection of saline solution. The vitamin D injection halted the progression of severe arthritis in the mice. These results suggest that vitamin D may hold promise for the control of arthritic lesions in Lyme disease. Further investigation, including an analysis of the blood serum vitamin D levels of human Lyme disease patients, has been suggested. To date, this research has not been expanded into human models. [94]

Another area of Lyme disease research has looked at risk factors for developing the disease. A meta-analysis of the results of previous studies done to assess the level of risk for Lyme disease among people who work outdoors in endemic areas was performed. After reviewing fifteen studies that were done on a worldwide basis, the report concluded that working outside increased an individuals chance of being exposed to the Lyme bacteria and testing blood seropositive. In the few cases where seropositivity was matched with symptoms, however, it was found that working outdoors in the sun had some protective quality. In one study in Spain, for example, thirty six percent of farmers in an area were seropositive for Lyme disease but none of them had any symptoms. Again, this might be explained by vitamin D blood serum levels. [95]

In a study of risk factors for Lyme disease in a small rural community in northern California, fishing, camping, farming, hunting, and hiking were found to be statistically protective in a variety of studies, while spending more time indoors was a risk factor.[96] Although they were not specifically testing for vitamin D levels, these results tend to confirm that time spent outside in the sunlight seems to protect people from developing the symptoms of Lyme disease. Time spent indoors may keep you away from ticks but if you get infected, an indoor life style may also increase your risk for symptomatic disease.[97] These studies have important implications for future Lyme disease research. There is a critical need to see if the protective qualities of vitamin D found in the mouse-model will translate into improvements in the human arena. Public Health policies that provide mechanisms for routine monitoring and improvement in vitamin D levels may be imperative. [98]

In the United States, federal funding for West Nile Virus research, which has had a minuscule death rate, has in recent years outstripped that dedicated to Lyme disease, and Bird Flu research for an anticipated pandemic that has not yet afflicted a single American has been very well funded. Vitamin D research has received very little funding even though scientific studies have begun to uncover a high level of vitamin D deficiencies in the American population. While millions of dollars have been spent developing problematic vaccines and other medications with high profit potentials, not enough attention has been paid to an improved vitamin D supplementation program. Supplementation may provide an

# AFFLICTION

inexpensive alternative to the health care costs of an ever-growing list of afflictions that appear to be vitamin D modulated.

If a human has been exposed to the *Bb* bacteria through a tick bite and has then gone on to become symptomatic, one of the earliest signs of infection can be the appearance of the Erythema Chromium Migrans (ECM) rash. This is usually described as the classic jagged edged round bull's-eye rash at the site of the tick bite, although homogeneously red rashes are more common, and not everyone develops a rash. Other signs of early infection are nonspecific: sore throat, fever, muscle pain, joint pain, palpitations, swellings and fatigue. These symptoms can appear suddenly. Within days to weeks of the initial infection, the spirochetes may disseminate to tissues throughout the body through vascular or lymphatic channels.

Lyme disease is an extremely frustrating ailment to describe because it presents such a diverse variety of symptoms. Because so many body systems can become involved, Lyme disease can take on a wide array of forms. In a very small number of people, it is an extremely severe pathology and can be fatal, especially if the heart muscle is infected or pneumonia develops. However, it is believed that prior to the establishment of an antibiotic protocol for Lyme disease it was generally a mild and occasionally a miserable but, in normal circumstances, survivable condition.[99]

In one sad study, ninety patients in the former Soviet Union with ECM and flu-like symptoms that developed after documented tick bites were studied, *but not treated with antibiotics*. Within two weeks to four months of infection, sixty-four percent of the patients had developed neurological abnormalities, including cranial neuritis, Bell's palsy, meningitis, or radicular pain. Four percent developed arthritis. In another study of patients with Lyme meningitis, it was found that, without antibiotics, the duration of neurological involvement was three to eighteen months. Untreated Lyme disease may have been responsible for some of the rheumatic conditions, aches, pains, dementia and mental illness that were sometimes thought to be a normal part of the aging process in the past in endemic areas.[100] This was especially true in New England where joint pain, "rheumatism" and quirky or eccentric behavior in the aged became an accepted part of human culture once witches stopped being a culturally accepted feature and witchcraft accusations diminished.

This diversity of symptoms may be in and of itself diagnostic of the affliction. Dr. Brian Fallon, the director of the Lyme and Tick-borne Diseases Research Center at Columbia University Medical Center, has suggested that if symptoms involving more than one specific bodily system are present, Lyme disease should at the very least be considered. "There are not many things that can cause brain problems and joint problems at the same time," he said.[101]

Although it is still the subject of much debate, the only diseases with strong evidence for a post 1492 New World to European transfer route are the spirochetal treponematosis diseases of syphilis *(Treponema pallidum)* and yaws *(Treponema pertenue)*. Lyme Disease *(Borrelia burgdorferi)* is a member of the same spirochetal family, which adds some validity to the suggestion of a post-1492 transfer of *Bb* back to continental Europe, especially to those places controlled or interacting with the Spanish Empire.[102] The *Bb* bacteria itself has a complex relapsing or shape shifting life cycle depending on its environmental circumstances.

Figure 11. *Microscopic view of Bb spirochetes.* CDC

It can change form. When the spirochetal form encounters adverse conditions like an immune system attack, an antibiotic, or even a dunk into distilled water, it can transform itself. It can change into a cell wall deficient, or spheroblast L-form, that can invade the host animal's own cells, spewing outer surface proteins (Osp) in the process, or into a cyst with a thick outer shell as a survival mechanism to escape unfavorable conditions. The cyst plays a role in protecting the bacteria's DNA with its genetic markers. Dr. W. Burgdorfer states that this cystic material is found in every animal and human tissue that is infected with *Bb*.[103] These cysts are highly resistant to destruction.

Some hormonal responses, however, especially those associated with menstruation, stress, or a weakened host immune system, seem to trigger an "all clear to revert to spirochete again" signal. So can the arrival of a new set of bacteria from a subsequent tick bite.[104] One recent study found that *Bb* spirochetes participate in a form of mating that transfers genetic material in the process.[105] Previously cysted *Bb* within a host's body may sense the outer surface proteins of newly arrived bacteria and transform into spirochetes to meet, mix and mingle with the newly arrived. This may be one of the problems discovered during the Lymerix vaccine test trials- people who had a pre-existing inactive cysted form of the bacteria within their bodies had it transformed in the process into an active spirochete infection by the injection of the new outer surface protein factor that the vaccine was made from, making them symptomatic. In people with the HLA genetic sensitivity, the new outer surface proteins in the vaccine may have triggered an overzealous immune response that caused damage to joints.[106] Researchers have found that the *Bb* bacterium contains a glycolipid that triggers an immune response from the body's natural killer T cells. The Lyme glycolipid is one of the few that naturally induce this immune response from T cells.[107] But there also appears to be the coexisting issue of *Bb* *causing* immunosuppression in some individuals. This mechanism may underlay the cases of chronic post Lyme affliction affliction that present a wide variety of symptoms: with a suppressed immune system whatever preexisting bacterial, fungal or viruses infections a patient carried within that had been held at bay by a healthy immune system would be free to wreak whatever variety of havoc it was capable of producing. [109]

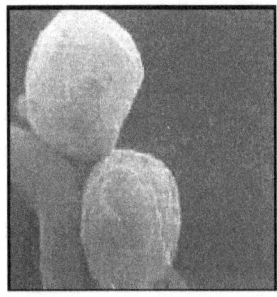

Figure 12. *Cyst form Bb*[108]

These bacteria have many tricks up their sleeves. Researchers at the University of North Carolina have found that during feeding, the *Ixodes* tick vector actually transmits a highly variable population of *Bb* into the host. In contrast, when the tick is not feeding, the population is fairly homogenous. What this means is that the bacteria essentially adapts during the transmission process to maximize its chance of infecting a blood meal host. Intriguingly, the fact that the tick spits so many different "flavors" of *Bb* into the host seems to explain an observation made several years ago: Lyme bacteria delivered by ticks evade the host's immune response more successfully than do cultured bacteria injected into animals. The team suggests that future research efforts could either focus on antigens produced within the tick before the bacteria population diversifies, or on surface proteins common to all of the otherwise variable bacteria. [110] A vaccine that causes the formation of Anti Tick Saliva Antibodies [ATSA] is another suggestion.

## DISGUISED AS THE DEVIL

Signs that a Lyme infection has disseminated (spread through the body) can include pulmonary, cardiac[111], neurological and musculoskeletal symptoms, headache, malaise, fatigue, profuse sweating, swollen glands and hoarseness, peripheral neuropathy, tremor, muscle twitching, vision problems including photophobia, severe sensitivity to sound and vibration, seizures, severe startle reaction, short term memory loss, sleep disturbance, hallucinations, and nausea or vomiting. In one study, patients who were bitten on the head or neck had a forty percent greater chance of having neurological problems. Lyme-associated neurological conditions that can sometimes develop are called cranial palsies. The most common (occurring in five to ten percent of patients) of these is Bell's palsy (which affects the face causing paralysis on one or both sides of the face.) Meningitis and encephalitis, inflammations of the membrane that covers the spinal cord and brain, and encephalomyelis, an inflammation of the brain itself, can occur in Lyme disease. [112]Because of this brain involvement, Lyme disease is also linked to psychiatric manifestations. Some patients have committed suicide. [113]

Although it is thought to be a rare occurrence, the *Bb* bacterium has been found to be secreted in milk, urine, and semen. It therefore may have additional pathways of infection that include sexual and congenital transfer although this is still subject to further research and controversy.[114] Modern studies of pregnant women from endemic areas in Italy who unfortunately suffered from undetected Lyme disease during their pregnancies show that spontaneous abortion, stillbirth and extremely rare cases of fetal deformity can occur.[115]One of the side effects found in the FDA trials for the Lymerix vaccine was a measurable rate of spontaneous abortion among pregnant women after inoculation.[116]

Modern veterinary studies have shown that domesticated animals are subject to Lyme disease infection. *Bb* spirochetes have been cultured from dogs, cats, goats, cows and horses. Cows that are infected can become lame and lethargic. They are also subject to a higher than normal rate of bovine abortion, may stop producing milk, or may pass the spirochetes along in their milk (it is killed by the heat in the pasteurization process).[117] Dr. Stephanie Holzman, a modern veterinarian, sees dogs and horses with Lyme disease that have "all sorts of symptoms or lack of symptoms," she said. If a dog comes in with kidney failure it will usually die. But if it comes in limping because of an affected joint, after being treated with doxycycline, "they'll never look back…" Holzman stated that "one of the hot topics at conferences is why we (vets)

# AFFLICTION

see so much variation in symptoms."[118] Another vet confirms the prevalence of symptoms in dogs adding, "of the hundreds of cases of canine Lyme Disease that I have seen, over ninety percent of canine patients were admitted with signs of limping (usually one foreleg), lymph node swelling in the affected limb, and a temperature of 103 degrees or more. The limping usually progresses over three to four days from mild and barely noticeable to complete disuse of the painful leg. Once the dog starts to be affected by the bacteria, Lyme disease can progress from a mild discomfort to the stage where a dog will be in such joint and muscle pain it will refuse to move; it is not uncommon for an owner to have to carry a sick dog into the animal hospital. Over the span of two or three days a dog can progress from normal to completely unable to walk due to generalized joint pain." [119]

Figure 13. *Peromyscus leucopus*

In the twenty-first century, Lyme disease is endemic in New England, epidemic in some areas, and Cape Cod has one of the highest incidence rates in Massachusetts. Was Lyme disease present in seventeenth-century Massachusetts? There are many tantalizing clues that push the disease's existence back to the seventeenth century and even earlier. The tick itself has an evolutionary history that spans at least 120 million years. The earliest fossilized tick from North America, *Carios jerseyi*, dated at 90 to 95 million years old, was found in a piece of amber excavated in central New Jersey. It was found in association with a small bird feather, which may denote the early precedent for the relationship between ticks and birds.[120] The subgenus *Ixodes* may have formed +/- 80 million years ago, prior to the break up of the proto-continent Laurasia into North America and Eurasia.[121]

The favorite blood meal hosts for *Ixodes* ticks in North America, white tailed deer, *Odocoileus virginiansis,* and white footed mice, *Peromyscus leucopus,* are also old species. Fossils have been found of the ancestors of the white footed mouse that date from the Oligocene Epoch of +/- 40 million years ago.[122] White tailed deer have a body structure that has not changed much since it speciated 3.4 to 5.2 million years ago. As

Figure 14. *Odocoileus virginiansis.*

generalists, whitetail deer are good survivors. When the environment is shaken up, whether by fire, earthquake, war, hunting, an Ice Age, or human development, whitetail out-survived their competitors whose more specialized needs may be less likely to be met in a changeable environment.[123]

*Borrelia burgdorferi,* the Lyme disease bacteria, has been traced by DNA amplification back at least to the Paleolithic Era of 400,000 to 15,000 years ago. It was first acknowledged as a human pathogen in medical literature in the United States between 1982 and 1983. The most studied transfer mechanism for Lyme disease is the zooinosis model, which involves the bacterium being present in local vertebrate reservoir hosts (white footed mice, migratory birds) and then being transferred incidentally to humans via an arthropod vector (tick). Lyme disease has been found to be endemic on a worldwide basis, depending on environmental conditions, caused by several different genetically distinct strains of *Borrelia* carried by various members of the *Ixodes* tick species.[124]

This pan-endemic occurrence and its complex shape shifting physiology are probably the strongest evidence to support a lengthy and protracted evolutionary existence for the bacteria. Over time the *Bb* bacteria has developed effective mechanisms to evade the immune systems of its blood meal hosts to survive and sometimes thrive. One DNA sequencing study has led to a hypothesis that the *Borrelia* pathogen is distantly derivative from African Swine Fever, which has a similar relapsing life cycle, mammal host vector (pigs), and arthropod (louse) mechanism for spreading.[125] In West Africa, there is a long history for *tick fever*, that is spread by mammal vector hosts (moles, rats and mice) and (*Ornithodoros sonrai*) ticks. After being eradicated in the 1950's, tick fever is reemerging as a major public health threat in the northwestern African countries of Senegal, Mali and Mauritania due to severe drought.[126]

European ticks were probably cross-infected from several points of contact over a long period of time including the crusades, after the Viking interactions with North America (Vinland), and again after 1492. The sixteenth century prevalence of a form of land scurvy in England and the English claim that this affliction was brought over by the Normans in 1066 may simply point to a dietary downturn for the conquered English people. But it could just as easily be the result of a tick and *Borrelia* bacteria accompaniment to the invading army, perhaps on or within humans, horses, dogs, cattle and swine.[127] A subsequent invasion of England by the French army that accompanied the return of Henry Tudor

AFFLICTION

in 1485 seems to have inaugurated an ongoing set of epidemics of a particularly virulent disease called *The English Sweating Fever*, which had some features that are similar to *Borrelia* induced relapsing fever.[128] However they got there, a wide variety of subspecies of *Borrelia* are found in various parts of Europe today.

The well-traveled Vikings may have been active players in Lyme disease dissemination between North America and Northern Europe, especially England. In the Icelandic Sagas, the areas around Vinland (thought to be a section of Newfoundland and the St. Lawrence River Valley in modern Canada) were described to be heavily populated by "dyr." The Vikings used puffins and other seabirds as a significant dietary resource.[129] Modern puffins that are found in the area between the islands off the coast of Maine in North America and Scandinavia in Europe have been found to be infected by *Ixodes uriae* ticks and to carry a form of the Lyme spirochete. Modern Scandinavia is an area where Lyme disease is prevalent.[130] It may have been prevalent in the past. Sweden also experienced a wave of disease and reactive witchcraft prosecutions immediately after the return of many of their soldiers from mainland Europe at the end of the Thirty Years War in 1648.[131]

Possibly the earliest written description of Lyme disease in Europe comes from the Island of Jura,[132] located off the coast of Scotland. This island was once within the Viking sphere of influence. After visits in 1764 and 1771 by the Reverend Dr. John Walker, he wrote "in the island of Jura, the cripples are remarkably numerous; owing to a very singular disease with which this island is infected." He goes on to describe a worm that is "of a reddish colour and of a compressed shape with a row of feet on each side" that "penetrates the skin with several small ichorous orifices [tick bites]…the worm disappears soon after this stage of the disease [a tick falls off after a blood meal] but when it is suffered to come this length, it never fails to cripple the patient for life. And the intense pain with which it is accompanied sometimes destroys the appetite and spirits and occasions death."[133] Some of the emigrants that came to New England and Pennsylvania during the early eighteenth century mass migration of the Scots-Irish originally came from Jura, perhaps bringing along a few hitchhiking ticks with their belongings. [134]

There are several different interpretations for the history of the North American spread of Lyme disease that can be pieced together. It may predate human contact. Endemic in Siberia, it may also have been carried over from Asia with the earliest human occupants and their dogs when they walked across the Bering Strait. The bacterial infection of

## DISGUISED AS THE DEVIL

Lyme disease may be detected by the appearance of scar damage to the bones of infected animals, especially in the joint areas. Lyme-like peripheral polyarthritis lesions (scars from infectious damage to the ends of multiple bones) have been found on one set of human skeletal remains in Alabama that date to 5,000 B.C. and there is such a large cluster of pre-contact skeletons with infected joints in the Tennessee area that it has been labeled 'the Cradle of Rheumatoid Arthritis.'[135] Another possible diagnosis of Lyme disease as the cause of similar pathological bone conditions comes from a prehistoric Tchefuncte Indian adolescent skeleton that dates from between 500 and 300 A.D.[136] This theme can be continued by theorizing that Columbus, landing on islands that were located on tick infested bird migration routes (including an astounding estimated population of five million of the now extinct Passenger pigeons) then brought Lyme disease [along with syphilis] back to co-infect Europe again from the New World after 1492. It is of some interest to note that Columbus himself suffered from severe arthritis, periods of temporary blindness, and hallucinations (all symptoms of late stage Lyme disease) only after he returned from the New World. He was so terrified during his fourth voyage of exploration by "a sea turned to blood boiling as a cauldron on a mighty fire" that he abandoned his ship for an extended period.[137] One of the earliest outbreaks of witchcraft accusations after Columbian contact occurred was on the Island of Barbados.[138]

Birds have been found to play a crucial role in both the spread of the disease in the Midwest and along the Atlantic coast of North America. They are competent maintenance hosts with good transmission capacities for the *Bb* spirochetes. Birds can be sensitive to temperature related ecotone variations in North America. Their ranges have been found to expand and contract as ecotones are displaced. However, population density and competition also play an important role in determining bird distribution. Infected birds play a prominent role in transferring *Bb* to larval and nymphal ticks as well as by transporting ticks into wide ranging areas.[139] Ground feeding birds, including robins and the common blackbird were found to carry ticks in a study done in Europe and to be competent reservoir hosts for the *Bb* spirochete in America.[140]

The first English settlers in Virginia arrived in 1607 and founded Jamestown in a swampy overgrown area. Almost immediately they began to suffer from what they called "the summer sickness" which included the symptoms of "swellings," fevers, extreme lethargy, and irritability.[141] All these symptoms could be attributed partially to Lyme disease. The

# AFFLICTION

Jamestown area is located along the major Atlantic coastal bird migratory route and is, in modern times, endemic for Lyme disease. The islands that Columbus first interacted with in 1492 are also located along these migratory routes. Birds continue to spread ticks and the infection along a swath of the North American ocean and lake coastline.

Edward Winslow, visiting a Native American village for the first time during the summer of 1621 and was plagued by a variety of vermin. He wrote, "lice and fleas within doors, and mosquitoes without, we could hardly sleep all the time of our being there." Ticks were sometimes identified as a form of louse, particularly as a wood louse, in the seventeenth and early eighteenth century so this may be an early reference to ticks of some sort.[142]

While pigs are highly susceptible to a wide variety of infective pathogens, two sets of pig skeletal remains from a 'Big Dig' related archaeological site in Boston may show evidence for *Bb* infections in seventeenth-century Massachusetts. One bone, a scapula, contains signs of a massive bacterial infection. Another set of bones, an immature pig skeleton (aborted fetus), may be an example of the "caste young" that John Winthrop wrote about in his journal. These bones show a deformity on the distal half of a tibia that would have been the result of a massive infection delivered via the blood stream.[143]

Studies of journals that were written during historic periods of intense interactions (marching through and camp-outs) in what are now Lyme endemic landscapes also show numerous references to lingering ailments. In a 1685 petition for a land grant in return for their service during the Narrangasset Expedition (also known as King Phillip's War), the veterans wrote:

> We think we have reason to fear our days may be much shortened by our hard service in the war- from the pains and aches of our bodies, that we feel in our sinews, and lameness thereby taking hold of us much, especially in the spring and fall.[144]

Hiking across the landscape of New York state, *Travels in North America* author Peter Kalm, a Swedish botanist, wrote on a warm June day in 1749 that the woods were "abound with wood lice, which were extremely troublesome...scarcely any of us sit down but a whole army of them creep upon his clothes." He predicted that "this small, vile creature may, in the future, cause the inhabitants of this land great damage unless a method is discovered which will prevent it from increasing at such a shocking rate."[145]

## DISGUISED AS THE DEVIL

Veterans from the American Revolutionary War of 1776-1783 (some of whom fought battles in the wooded areas of New York that Peter Kalm had visited thirty years earlier) and the American Civil War of 1861-1865 suffered from high levels of subsequent "rheumatism." Records for Revolutionary War veterans are somewhat rare, but even those often list rheumatism contracted while in service as a disabling condition.[146] Even Andrew Jackson, the revered backcountry military leader who must have traversed a variety of tick risky landscapes in his day, was afflicted by severe rheumatism that he felt he had acquired during his military career. He was so ill just prior to the Battle of New Orleans that he could hardly stand![147]

Lewis and Clark's Corp of Discovery probably encountered ticks during their famous expedition when they passed through the ideal tick habitat of herbs and grasses, sagebrush, and juniper on their journey to the Pacific Ocean. William Clark may have suffered from the first recorded case of Colorado Tick Fever. For four days at Three Forks, Clark was prostate with weakness, fever, chills, and muscle pains after developing a painful swelling over the inner bone of an ankle that may have come from an unnoticed tick bite. The disease seems to have had a relapsing nature, because two weeks after recovering, Clark was struck with another illness that lasted eleven days- "I was again attacked with a violent Pain in the Sumock & bowels." In a letter dated February 25, 1804, Clark wrote:

> "I was at that time so unwell that I could not [write], my health at this time is Somewhat better thoh not entirely Recovered from the Indisposition which attacked me in assending the river.... I thought at one time I had entirely recovered, but haveing frequint returns of the disorder I am induced to believe that time and attention alone will destroy the effects [of] it."

Clark also recorded in his journals occasional remarks about rheumatism in the neck. Clark's experiences would be repeated in later overland travelers to the west who added graphic descriptions to the history of Colorado Tick Fever. William McBride, an 1850 emigrant, wrote:

> At dawn this morning E. W. Summy and Lewis Mitchell applied to me for medications. They were severely and suddenly attacked last night after partaking of a hearty supper. Their disease is what the Mormons have termed mountain fever, being a disease particular to this region. Many patients complain of a most violent pain in the head and eyes with dimness of vision, pain in the back and limbs, great lassitude, alternations of chills and fever, nausea and vomiting, constipation. Such are the most prominent symptoms. The pain from this disease is very great! I prescribed for them and this evening they are some easier. I saw a company of nine men today which have been compelled to lay at Green River a week, five of their men having the mountain fever at one time-it is very common at the river. The disease cannot be said to be very dangerous, but it is severe.[148]

Tick borne disease may also appear in records from the Civil War. A Civil War veteran census that was taken in 1890 showed that disability from rheumatism afflicted about twenty-five percent of that population,

# AFFLICTION

second only to bodily damage from gunshot wounds. In numerous instances, it states that men went into the army in full health but had suffered from rheumatic joint pain ever afterwards. One Civil War physician wrote about another symptom shown by some of the soldiers he treated. He called it Soldier's Heart, the most dominating symptom being profound fatigue.[149] It is possible that some of these ailments were actually Lyme disease contracted by men under the stressful conditions of combat while traversing, camping in, and otherwise intensely interacting with a mosaiced and "tick risky" landscape, especially the coastal area of Virginia and Maryland during the Peninsular Campaign.[150]

But by 1872, Asa Fitch, a naturalist who was retracing Peter Kalm's earlier travels, noted that in New York, the woodlands were gone and as a result, the ticks were "nearly or quite extinct." But the *Bb* was still present. An examination of museum specimens' DNA for the *Bb* bacteria has yielded an earliest found actual infection date in North America of 1894 for a white footed mouse from Dennis, Massachusetts (located on Cape Cod).[151] The earliest preserved *Borrelia* bacteria found in Europe sets an 1884 infection date for an *Ixodes ricinus* tick from a museum collection in Germany.[152]

Native American oral history can also be examined for evidence of a pre-contact North American existence of Lyme disease. There are at least two traditional stories that clearly link deer with arthritic conditions. In a traditional Creek story, hunters are warned to be cautious with deer because they have mysterious powers. If a hunter does not show proper respect when he has killed a deer, they would cause him rheumatism and the hunter would be forced to walk the rest of his life with aches and pains. Another Native tribe, the Cherokee, occupied parts of Appalachia that have been labeled as the 'Cradle of Rheumatoid Arthritis.'[153] Cherokee folklore contains both the strong admonition against touching the skin of a diseased deer and the following story about respecting the spirit of slain deer:

> Little Deer, the chief of the deer and the animal spirit who took vengeance on unthinking hunters, ran as swiftly as the wind to a deer just killed. Bending over the blood spots on the ground, Little Deer asked the spirit of the deer if it had heard the hunter make amends, the proper prayer. If the answer was yes, then Little Deer left. If no, Little Deer followed the trail of blood left by the hunter who then carried the deer to the hunter's door, where he "[put] into the hunters body the spirit of rheumatism that shall rack him with aches and pains from that time henceforth."[154]

There are litanies of Native American herbal treatments for arthritis and rheumatism showing that these were diseases that indigenous people confronted. Some of these Native cures, like bearberry and

## DISGUISED AS THE DEVIL

goldenseal, have mildly antibiotic properties that would possibly be somewhat effective for treating Lyme disease.[155]

Lyme disease is today recognized by the United States Center for Disease Control as "the leading cause of vector-borne infectious illness in the United States." [156]Lyme disease creates a variety of symptoms and its virulence can vary from year to year and from person to person. The experience of Lyme disease can be understood by looking at the medical and scientific literature. But it is also a very idiosyncratic condition that can be described either by the person experiencing the symptom or by an observer of the afflicted, based on his or her own senses. A growing body of modern source material can be found that includes descriptions of the experiences that Lyme disease victims feel. Many of these descriptions mirror information recorded during the Salem Witch Trials of 1692.

My own interest in this topic was piqued when I read, for the first time, descriptions of the pinching and the sensation that invisible needles were being pushed into the flesh of people who thought they were being afflicted by witchcraft. These descriptions accurately describe some of my own symptoms after contracting Lyme disease. To this day, I sometimes
feel as if someone is poking a hot spike into my shoulder and my skin often crawls and pricks for no apparent reason. Sometimes it feels like someone has tapped me on the shoulder but no one is there. I have learned to ignore many of these symptoms. I have also observed an acquaintance who suffered from Lyme-induced Bell's palsy. It is a terrifying sight even in the twentieth century when the cause is known, it must have been even more so in the past.

On a personal note, I contracted Lyme disease for the first time in November of 1994. At the time, my work involved creating historic programs for a living history farm. After working a long weekend of special events I went home feeling tired. Within a few days I woke up in the morning to find myself completely drained of energy-almost paralyzed. I could hardly get out of bed. My son, it turns out, was found with a tick attached to his leg and contracted the disease at almost the same time. Mistaken for a thorn on a squirmy toddler's leg, there was a delay in removing the tick. It was only when I tried to pull out the 'thorn' that I realized that it had legs and was alive. As the disease developed, his response was to howl with pain, especially when he bent his legs. Our pediatrician was reluctant to give him antibiotics at first because an early blood test came back negative but after we had spent three full days with his screaming in pain for most of the time she

# AFFLICTION

relented. The antibiotics made his pain go away and he seems to have been completely cured.

Antibiotics also helped me, but in the years that have followed, I have been forced to quit a job that I loved, and deal with symptoms that wax and wane. I often feel that I have had two lives- the one before Lyme disease and the one after Lyme disease, where nothing is quite the same. Whether my condition is called Chronic Lyme disease, Fibromyalgia, or Post Lyme Syndrome does not really matter to me- it is a life altered by the infection. And I am not alone. There are many victims of the chronic form of this debilitating disease.

There is a growing body of literature concerning the experiences of modern Lyme patients. Most of us live our lives in relative obscurity. But in some famous cases, the disease has played out across the pages of national
newspapers. President Bush has been infected, as have Senator Chuck Shumer of New York and former Governor Christy Whitman of New Jersey. Darryl Hall, of the singing group Hall and Oates, was forced to cancel a concert tour after contracting the disease.[157] The promising career of golfer Tim Herron was sidelined by severe fatigue and what felt like a sinus infection, cold sweats, and a full body rash in 2004. It took a string of emergency room visits and four doctor's opinions before he was diagnosed with Lyme disease. [158]

In June of 2005, Wyatt Sexton, a promising quarterback for the Florida State University football team, was found disheveled and disoriented lying on a street in Tallahassee. He identified himself as God and was taken to a hospital after being doused with pepper spray by the local police. At first, some form of drug abuse was suspected because these aberrant symptoms had appeared so rapidly. Eventually, he was found to have an active Lyme disease infection that had resulted in neuro-psychiatric and cardiovascular deficits.[159] As I was writing part of this chapter in the summer of 2006, Phil Bredesen, the Governor of Tennessee has contracted a "sudden acute illness" that may have been tick-bite related. He was sent to the Mayo Clinic for further testing.[160]
Polly Murray contracted Lyme disease over thirty years ago. She lived in Lyme, Connecticut where she, her husband, and three of her children may have been infected by the disease sometime before the early 1970's. Her work, along with many others, in trying to find answers to what was happening within her family and community, has been key to raising scientific and public awareness of this disease. Her book, *The Widening*

## DISGUISED AS THE DEVIL

*Circle,* is an autobiographical account of her own experiences with Lyme disease.[161]

Another well-documented modern experience with Lyme disease comes from the novelist Amy Tan, who was diagnosed with a case that had advanced into late stage symptoms. She has been very vocal about its devastating effects. She suffered from hallucinations (usually at night when she was in bed) which she has described. Once she saw a naked man approaching her bed and thought it was her husband. "It was the middle of the night," and "he wasn't saying anything or doing anything else. He was just coming toward me (before stopping) next to the bed stand, as though he was turning on the light." She "thought someone was dead. I reached for him and the image started to warp as I realized he wasn't real." The experience was terrifying. Amy Tan's other symptoms have included hair loss, fatigue, tinnitus, memory loss, olfactory hallucinations, and the misspelling of words when writing. When speaking, even after antibiotic treatment, she sometimes replaces words with similar sounding gibberish.[162]

Modern research has found that because so many body systems are involved, Lyme disease can take on a wide array of forms. People with mild infections may show no signs at all. In other people, Lyme disease is so severe that it completely disrupts normal life. Lyme disease can become chronic, or it can follow a cyclical pattern of active infection, remission, and relapse. Except for fatigue and lethargy, which are often constant, the signs of Lyme disease are typified by their intermittent and changing nature. Lyme meningitis is often accompanied by irritability, photophobia, and impaired concentration. Neurological abnormalities may last for months but will usually resolve completely even without antibiotic therapy.

Lyme disease can infect the brain. The relatively recent development of brain scans, especially the MRI and SPECT technologies has made the brain damage done by *Bb* spirochetes visible to researchers for the first time. In neurological Lyme disease MRI scans have found white matter hyper-intensities in the brains of approximately fifteen to forty five percent of patients. These seem to be similar to the damage seen in multiple sclerosis patients.[163]

Lyme disease has three types of skin manifestations that can appear: lymphocytoma, Acrodermatitis Chronica Atrophicans (ACA) and ECM. Lymphocytoma and ACA are prevalent in Europe but rarely occur in the United States. A lymphocytoma is a raised red lump that grows on various areas of the body. Anywhere from fifty to eighty percent (depending on which study you cite) of those who are bitten by infected *Ixodes* ticks develop a *bull's eye* rash or ECM, at the site of the tick bite. An ECM may be shaped like a ring, a triangle, an oval or a long thin ragged line and can be hot to the touch, itchy, burning or painful. Alternatively, the skin at the site can be numb.[164]

Did the lymphocytoma or this rash appear in the past? In his eloquently written book *The Biography of a Germ,* Arno Karlen mused that "it is understandable that people failed for a while to identify Lyme disease in all its complexity, but the ECM part, the bull's-eye rash of its first phase, is hard to miss. If Americans did not notice it before the 1970's, perhaps it wasn't here." [165]The answer to his quandary is that both the lymphocytoma and the ECM **were noticed** for centuries in both Europe and America. The ECM was called by several names, including 'the mark of the devil' or *'diablo stimata.'* The lymphocytoma was called a 'witches teat.' By the twentieth century, the existence of these marks of a witch had become so hidden under such a dusty folkloric layer that they were not noticed. But a mark on the skin was just as likely to appear in the records of a witch trial in the past as DNA is in a murder trial today.

Europe went through a series of climatic, economic and social shocks that eventually brought the medieval period to an end. In the early fourteenth century, 'The Black Death' [bubonic plague] killed off more than one third of the population in many areas. The arrival of the Little Ice Age heralded a colder climate. Even the prevailing definition of man's relationship with God and church came under attack with the Protestant Reformation. Over time, it became common knowledge that the devil himself had been let loose and was walking the earth causing great distress. According to witchcraft theories that became established in European thought after 1500, the devil sealed the compact he made when he created a witch by giving him or her some mark of identification on the skin- described as being not unlike the brand an owner might place on his cattle. This devil's compact and mark were considered by Christina Larner, a scholar of the Scottish witch trials, to be a uniquely European feature of witch beliefs. These ideas were carried to the new world during colonization. [166]

There is an entire body of literature describing this mark in words but it is rarely depicted. An illustration of the devil carrying a witch off to hell from Olaus Magnus' *Historia de Gentibus septentrionalibus* of 1555 may show one on a witch's foot. [See figures 15][167] One of the earliest writers to mention the mark was the Calvinist theologian, Lambert Daneau. In *A Dialogue of Witches*, he wrote that there was not a single witch, "upon

## DISGUISED AS THE DEVIL

whom [the devil] doth not set some note or token of his power and prerogative over them."

Judges in witch trials were advised that suspects should be carefully examined- "pull and shave, where occasion shall serve, all the body over, lest haply the mark may lurk under the hair in any place." [168]Sinistrari, another early demonologist, wrote: "The demon imprints on [the witch] some mark, especially on those whose constancy he suspects. That mark, however, is not always of the same shape or figure; sometimes it is the likeness of a hare, sometimes like a toad's foot, sometimes a spider, a puppy, a dormouse. It [can be] imprinted on the most secret parts of the body; with men, under the eyelids or perhaps under the armpits, or on the lips or shoulders, the anus, or elsewhere; with women, it is generally on the breasts or private parts. Now, the stamp which makes these marks is simply the devil's talon."[169] Finding such a mark was considered proof that a suspect was a witch and was in itself sufficient to justify torture or a sentence of death.[170]

Although searching the skin of accused witches for marks was delegated to a committee of women in Salem, even Cotton Mather commented on these marks: "I add, why should not witch marks be searched for? Divers weighty writers describe the properties, the qualities of those marks. I never saw any of those marks, but it is doubtless not impossible for a surgeon, when he sees them, to say what are magical."[171] The marks were described as being found on any part of the body. In England, they were often described as being on a finger.

If the mark was not immediately visible, it was suggested that it be looked for in more intimate hiding places. In *Disquisitionum Magicarum* of 1599, for example, Del Rio suggests that a search of the sexual parts and anus was often in order.[172] Because an ECM often appears at the site of tick attachment, parts of this searching of the body ritual continue in the modern day in the form of 'checking for ticks' which can be a very intimate examination indeed. It has permeated American popular culture. In the summer of 2007, my car radio often featured Brad Paisley's serenade:

> "I'd like to walk you through a field of wildflowers,
> And I'd like to check you for ticks....
> Ohh you never know where one might be"[173]

## AFFLICTION

The folkloric connection between the devil's mark and the intimate sexual areas often favored by ticks for attachment is often implicit in the epistemology of witchcraft. The witch offers the intimate parts of her body to the devil who then takes full advantage. At North Berwick in Scotland, court records described the initiation of witches as involving a sort of ceremony where "the devil doth lick them with his tongue in some privy part of their body, before he doth receive them to be his servants, which mark commonly is given them under the hair in some part of their body." When Agnes Sampson, an accused Scottish witch was shaved, "the devil's mark was found upon her privities." In 1658, another Scottish witch, Margaret Taylor, confessed that the devil, "gave her his mark . . . in her secret member."[174]

Dr. Jacques Fontaine wrote in *Des Marques des sorciers et de la réele possession que le diable prend sur le corps des hommes* of 1611, that "writers . . . who say that it is difficult to distinguish devil's marks

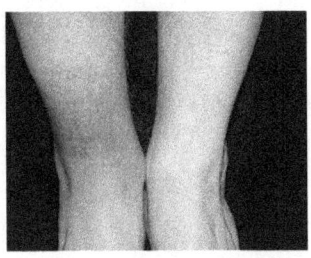

# DISGUISED AS THE DEVIL

**Figures 15.** *A modern ECM on the ankle www.hvcn.org [top left] can be contrasted with a devil's mark on the foot of the witch[bottom] shown in The Devil Carrying a Witch off to Hell, by Olaus Magnus, 1555, a woodcut illustration.*

from natural blemishes, from a carbuncle, or from impetigo, clearly show that they are not good doctors." Dr. Fontaine wrote in some detail about how the devil made these marks. "Some say that Satan makes these marks on them with a hot iron and a certain unguent which he applies under the skin of witches. Others say that the devil marks the witches with his finger, when he appears in human form or as a spirit. If it were done with a hot iron, it would necessarily follow that on the part so marked there would be a scar, but the witches testify that they have never seen a scar over the mark…. But it is not necessary to prove this, for the devil, who does not lack knowledge of medications and has the best of them, has only to mortify that place. As for the scar, the devil is such a skillful worker that he can place the hot iron on the body without causing any scar."[175]

Peter Ostermann, whose 1629 *Commentarius Juridicus* was a summary of both popular and official opinions at the height of the witch beliefs, wrote that "Not one person could be produced who, having the mark, had lived a blameless life; no one convicted of sorcery had ever been found without the mark. It was the proof of proofs and more infallible than either accusations or confessions."[176] Even Margaret Murray, whose 1921 work, *Witch Cult in Western Europe* appears to be more fantasy than fact, noted the persistence of historic references to the devil's mark. She theorized that it was a tattoo that was applied during the initiation ritual of a witch coven.[177]

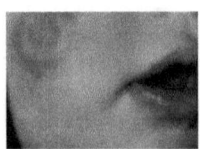 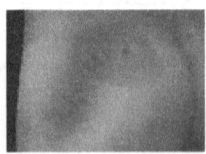

**Figures 16.** *Pictures of Lyme bull's eye rash on a buttock and a cheek., www.hvcn.org*[178]

Because they were associated with the epistemology of witchcraft, a deliberate search was made for marks on the skin, many people in the past appear to have noticed these marks, and their existence was consistently noted in writing in both Europe and America along with a coterie of explanatory theories. The ECM of Lyme disease should be added as an explanation for the devil's mark and marks found on the skin of those thought to be afflicted by witchcraft. The lymphocytoma, a raised bluish red skin lump, which is found in modern Lyme disease cases in Europe, may be the physical manifestation of the folkloric witch's teat concept that was brought to America. This bluish-red lump is usually found on the areas of the body that have low skin temperature like the earlobe, nipples and areolae, the nose, or the scrotum.[179]

In the historical record from the Salem witchcraft trials more items related to skin manifestations appear. Not only were accused witches searched for "witch's teats" but there are also numerous descriptions of skin manifestations appearing on the bodies of the afflicted. These ranged from a thin red line across a palm to a mark appearing that was round and resembled a bite mark made by the upper and lower teeth.[180] In an interesting epistemological turn, marks on the skin in early New England are as closely linked to those thought to be afflicted by witchcraft as they are to those accused of doing the afflicting.

The afflicted Goodwin children, for example, developed red streaks on their bodies, similar to the rash from the tick-bite induced co-infection of *Bartonella*. Another rash was described as being "the exact print, image and colour of an orange made on" a child's leg. In 1682, Mary Hortado of Berwick, Maine, had markings that looked like "the impressions of the teeth being like a man's teeth" that were plainly seen by many and also developed a set of red blisters that looked like "scalds" from a heat source. These marks would develop and then disappear within a few days. In 1662, the daughter of John Kelley died with a large tawny spot on her cheek.[181]

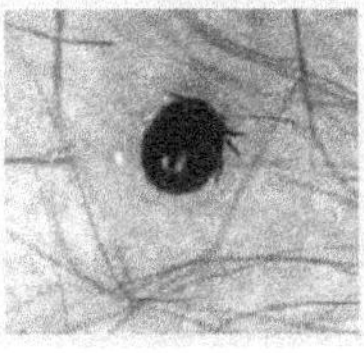

Margaret Rule was observed to have blisters "raised on her skin" which were treated with the "oyl's proper for common burning" and that most of the scalds also went away without scabbing up or scarring.[182] In 1685, Jarvis Ring of Salisbury developed "the print of the bite (of a woman)" on the little finger of his right hand. [183]Deodat Lawson, a minister visiting Salem Village during 1692, observed that the afflicted Mary Walcut., "as she stood by the door, was bitten, so that she cried out of her wrist, and looking on it with a

Figure 17. An attached engorged Ixodes tick has the appearance of the head of a 17th C. pin. CDC

candle, we saw apparently the marks of teeth both upper and lower set, on each side of her wrist." An undesignated number of the afflicted "produced the marks of a small set of teeth." In proper English tradition, even four-year-old Dorcas Good, an accused witch, was observed to have a "deep red spot, about the bigness of a flea bite" on her finger. [184]

The devil's mark is only one of many clues about the existence of Lyme disease that can be found in the past. Because "The past is a foreign country: they do things differently there" it is important to try to understand these differences. For example, some people observed pins that seem to appear out of nowhere and became stuck into the flesh of the afflicted. The modern temptation is to ridicule these observations as fantastical. But things were indeed different in the past. Even something as minor as a pin (which plays a prominent role in witchcraft in both Europe and America) had a very different appearance in the seventeenth century. Most pins and needles today are made of shiny steel. Although needles sometimes have the traditional brass colored designation for the eye end (an eons-old persistent remnant from the distant past when poor vision was not yet correctable) it is difficult to imagine mistaking a modern pin for an attached tick.

Although they are a somewhat rare archaeological find, in the seventeenth century pins had a different appearance. They were usually handmade of a dark colored brass or iron with a head made by pounding a brass wire twisted around the top until it was attached.[185] This created a bulbous brass colored circular top that often had a round center of darker metal. While a shiny modern pin would *never* be mistaken for an attached tick, what would happen if the visual image of an embedded pin closely mimicked both the appearance and colors of an attached adult deer tick? In the past it did. A nymphal *Ixodes scapularis* tick has a decidedly brassy coloring. A female adult is bi-colored with a brass or rust colored body and a dark brown or black mouth end area with a clear circular delineation between the two. When attached, a tick is not only stuck into the flesh, it may contract its legs and can assume a pin or thorn like appearance.[186]

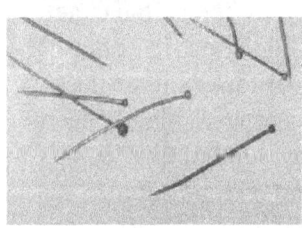

**Figure 18.** *Iron & Brass 17th C. Pins.*
*Collection of Author*

When Mercy Short lay down on a bed in Boston in 1692, and, to the horror of onlookers, pins were later found sticking into her flesh, could they have been attached ticks? Had she come to acquire *Ixodes* ticks when she visited the jail that housed several accused witches

# AFFLICTION

from the Salem Village area? Were ticks inadvertently brought into the jail in Boston on the skirt hems of accused witches? The floor of the jail was covered with a layer of sawdust or wood shavings, which may have been moist with water and urine. This might have made an excellent microhabitat for *Ixodes scapularis* ticks right in the heart of Boston. We know that Mercy reached down and picked up a handful of these wood shavings to throw at the accused witch, Sarah Good, who had begged for tobacco. Did that action bring her into closer contact with ticks?

Alternatively, did the bed she was lying on, like others in the seventeenth century, have a mattress that was stuffed with wood shavings, straw, cornhusks, eelgrass or even dry leaves? Was this mattress infected with bedbugs that included ticks? While the preternatural pins fit the seventeenth century mental mindset of witchcraft, there is also a plausible modern explanation that the pins that were observed were actually attached ticks. Mercy Short could also have been re-infected. She had been forcibly marched from Maine to Canada as an Indian captive during an earlier battle and had been rescued during the attack on Quebec made by the Phips expedition. It is likely that the stress of that situation may have included contact with ticks.[187] Many of the pins mentioned in the New England

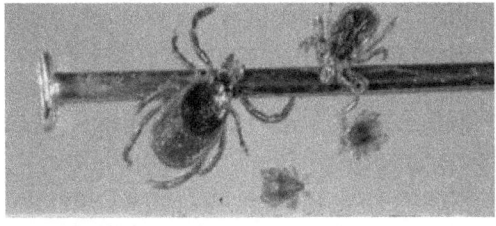

**Figure 19**. *Enlarged view of ticks next to modern pin.* CDC

witchcraft trial cases are, however, invisible-*felt but not seen*-but they are definitely part of the etiology of witchcraft affliction. It should be noted that the sensation of peripheral neuropathy is commonly called a feeling of pins and needles.

Cotton Mather described these pins as a mixture of seen and unseen pins that invisible but "barbarous visitants" of some sort could insert and pluck out again (after taking a blood meal the tick detaches and falls off) leaving behind bloody marks which were cured rapidly. Some pins, however, were "left in" Mercy Short and were removed by onlookers with "wonderment." In 1684, Philip Smith of Hadley also had an encounter with pins. According to Cotton Mather, he "cryed out not only of sore pain, but also of sharp pins; sometimes in his toe, sometimes in his arm, as if there had been hundreds of them." Upon search of his body, one pin was found attached.[188] Witch Trial critic and skeptic Robert Calef himself saw these pins and wondered 'what sort of Distemper 'tis

## DISGUISED AS THE DEVIL

shall stick the body full of pins, without any hand that could be seen to stick them?" Writing about the afflicted Margaret Rule, Calef states that she would every now and then be miserably hurt with pins which were found stuck into her neck, back and arms, however, the wounds made by the pins would in few moments ordinarily be cured."[189]

Looking for signs of the devil's doing, people observed pins and marks on the skin and the descriptions are dead ringers for attached ticks or the characteristic bull's eye or ECM rash of Lyme disease. They experienced swellings and pain in their joints that was often accompanied by neurological symptoms. A baby boy in the 1680's Salem suffered from what might be diagnosed as Bell's Palsy[190]- a possible clue that the Lyme bacteria may have infected his cranial nerves. People suffered severe fatigue, had seizures, and were terrified by nightmares and hallucinations. Flu-like symptoms abound in the literature. The afflicted Godwin children suffered from fevers and chills. "They would sometimes complain that they were in a red-hot oven, sweating and panting" and then feel that "cold water was thrown upon them, at which they would shiver very much." The afflicted sometimes suffered from chronic sore throats, extreme fatigue and would lie in bed "for diverse weeks" and have fainting and swooning spells afterwards. Ann Putnam, Sr. suffered from extreme fatigue and was barely able to walk at times.[191] A badly swollen gland is described as "a ball as big as a small egg, into the side of her wind pipe."[192]Even Cotton Mather's first wife, Abigail, after giving birth to a deformed baby suffered from a lingering sore throat and chronic ailments on and off for years.[193] The Lyme-like sporadic nature of symptoms and outcomes was noted by many of those writing about the afflicted in the seventeenth century. Thomas Brattle wrote that "many of these afflicted persons, who have scores of strange fitts in a day, yet in the intervals of time are hale and hearty, robust and lusty, as tho' nothing had afflicted them." One of the afflicted became severely obese. Elizabeth Knapp, whose illness lasted for nearly three months between October 30, 1671 and January 15, 1672, was "no ways wasted in body or strength by all these fits, though so dreadful, but gathered flesh exceedingly."[194] This parallels the modern experiences of feeling dreadful but not *"looking"* sick and the fact that many women

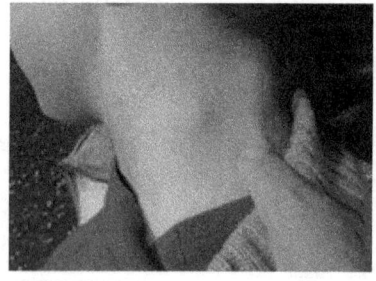

Figure 20. *A swollen gland of Rickettsia.* CDC

# AFFLICTION

with Chronic Lyme tend to gain excessive weight. Cotton Mather notes the fact that symptoms seemed to wax and wane in both the afflicted Goodwin children and Mercy Short. Symptoms would sometimes disappear completely only to return after some weeks of health. Modern research has found that infected women may have more risk for experiencing cyclical periods of more difficult Lyme symptoms because their immune system functions are influenced by menstrual hormone changes.[195] The vast majority of the afflicted from 1692 were female.

It is the neurological component of Lyme disease that, when combined with the observation of pins and skin rashes or marks, makes it a compelling diagnosis for those who were afflicted by what they felt were preternatural forces. When the mind itself is affected, the Lyme sufferer must deal with the odd nature of some symptoms. The sensation of a poke on the back, pricking on the skin, or the shooting sensations of heat are experienced as real sensations. The mind interprets a hallucination as a very real phenomenon. The very real physical response of seizure, sweat, tension, and hormonal release can be triggered by this bacterial infection. Some of these symptoms had appeared so regularly that by the beginning of the seventeenth century a sixty seven item list of diagnostic symptoms useful " to distinguish demoniacs ...from the simply bewitched" in Europe was included in the 1608 guide *Compendium Maleficarum.* With minor revisions, this list could be used as a modern physician's guide to the myriad of symptoms of untreated Lyme disease.[196]

Cranial nerve palsies can be Lyme associated neurological conditions. Understanding these palsies is important to this work because they may have caused many of the symptoms described or experienced by *people* in *seventeenth century* New England *(see Appendix G for descriptions of some symptoms)* as noted in the historical literature. There are twelve pairs of cranial nerves that can be affected by Lyme disease and create specific symptoms, including:

Cranial Nerve Pair I- Olfactory: afflicted may experience loss of smell or smells may be overly intense or noxious.

Cranial Nerve Pair II- Optic: Partial or total loss of vision may occur. *This symptom was exhibited by the Goodwin children and the Howen boy, Benjamin Holten, John Parker, Josiah Eaton.*

Cranial Nerve Pair III- Oculomotor: The eyelids may droop, the eyeball may deviate outwards, or the pupils may become dilated. Some patients may squint involuntarily or see double images.

Cranial Nerve Pair IV- Trochlear: The eyeball may rotate upwards and outwards or double vision may occur when looking down. *Mercy Short*

## DISGUISED AS THE DEVIL

Cranial Nerve Pair V- Trigeminal: Pain or numbness in parts of the face, scalp, forehead, temple, jaw, eye, or teeth has been reported. The muscles

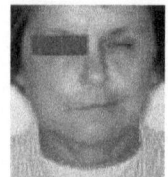

Figure 21. *BELL'S PALSY, which affects Cranial Nerve Pair VII creates paralysis on one side of the face. That side may sag, the eye may not open, and the mouth may have a leering appearance-t preternatural grimace. This palsy is one of the known manifestations of neurological Lyme disease and can come and go.* CDC PHIL

used for chewing may become paralyzed or dysfunctional, making it difficult to chew, and the jaw may deviate towards the paralyzed side. *Mercy Lewis, Margaret Rule*

Cranial Nerve Pair VI-Abducens: The eye may deviate outwards, and excessive squinting or double vision may occur.

Cranial Nerve Pair VII-Facial: The improper functioning of these nerves can result in Bell's palsy on one or both sides of the face. Hearing loss can occur on the affected side. Problems chewing and speaking may result. *The Shattuck children*

Cranial Nerve Pair VIII-Vestubulocochlear: hearing disturbances such as tinnitus, full or partial hearing loss can occur. Equilibrium problems can occur causing dizziness and falling down. The *Shattuck son, Phoebe Chandler, the Goodwin children*

Cranial Nerve Pair IX-Glossopharyneal: Problems occur with the mucus membranes in the back third of the tongue, may cause difficulty in swallowing and paralysis of the glottis.

Cranial Nerve Pair X-Glossopharyngeal: Dysfunction of the muscles in the throat, shoulders and back may cause difficulty in swallowing or talking. Drooping shoulders and an inability to rotate the head away from the dysfunctional side may also occur. Malfunctions within these branches of this nerve may cause heart problems, breathing difficulties (including slow respiration and a sense of suffocation), cough, glottis paralysis vocal cord spasms or paralysis and gastrointestinal problems. *Mary Hortado, Morse boy, a Shattuck child, John Louder, Jarvis Ring, Phoebe Chandler, John Parker, Josiah Eaton, Stephen Bittford, the Goodwin children.*

Cranial Nerve Pair XI- Spinal Accessory: Disrupted function or paralysis of the upper back and neck. Back spasms and the inability to tilt the head to the shoulder or rotate the head in either direction can occur.

Cranial Nerve Pair XII-Hypoglossal: Affects the tongue, speech and swallowing. Can cause a deepening in the tone of the voice. *Elizabeth*

# AFFLICTION

*Knapp's tongue was drawn out of her mouth and spoke in a voice that was not her own, Mercy Short, the Goodwin children, the Morse boy all had tongues either drawn in or hanging out to great length.*

Meningitis, an inflammation of the membrane that covers the brain and, encephalomyelis, an inflammation of the brain, can occur in late stage Lyme disease. Encephalopathy, a dysfunction of the brain itself, can also occur. Cognitive problems and mood disturbances such as irritability, bursts of crying, and temper flares occur. Profound fatigue, confusion and difficulty sleeping are reported. Depression, mood changes, hallucinations, panic attacks, Tourette syndrome, verbal aggressiveness, extreme agitation, inappropriate and manic behavior and psychosis have been reported. Some patients have attempted suicide.[197]

Lyme disease can be accompanied by seizures. The mother of a teenager from an endemic area of New Jersey described the seizures that he developed after contracting Lyme disease. He had "episodes of tremors, wherein his body would shake and twitch uncontrollably."[198]

During the Salem Witch Trials, oral testimony was written down, although it was often by one person-the Rev. Samuel Parris, who can hardly be described as an uninterested bystander.[199] While there seems to be some selectivity in terms of what records remain and which have been intentionally destroyed over time, the records that do remain provide a rare opportunity to study the afflictions suffered by members of a community from the past. Those who are usually- because of time, space, gender and insignificant social standing rendered inarticulate- are instead given loud voices. Although records are fragmentary at best, Salem Village of 1692 remains one of the most articulate communities from America's collective past. Each person called to testify is given a forum to describe the world from his or her own point of view.

Everything from the mundane acts of yoking a pig or conversing with a neighbor to the extraordinary sensation of flying over the tree tops in the dead of night or the amazement at the combustibility of a mare's flatulence are included. A set of beliefs emerges that firmly include the existence of witchcraft as a cultural construction-- a way to understand aberrant behavior, disease, and death.

Modern scholars have to a large extent relegated the "afflictions" of 1692 to an imaginary world somewhere between the psychiatric ward of history and a special hell reserved for liars, belittling the "girls'" afflictions, judging their veracity, and preferring to concentrate on more important parts of the story. While no one has put the episode into an

## DISGUISED AS THE DEVIL

environmental context, there are many excellent books written about the Salem community. Many pre-date the modern discovery of Lyme disease. John Demos wrote a description of Salem's place in Puritan social history.[200] Stephen Nissenbaum and Paul Boyer's *Salem Possessed* outlined the social and political factions of Salem.[201] Mary Beth Norton's recent *In the Devil's Snare* added an Indian Wars posttraumatic stress psychological evaluation to the body of work.[202] However, in only a few instances has any attempt been made to analyze the "afflictions" of the accusers from Salem Village as anything other than imaginary and find possible causes.

Chadwick Hansen in *Witchcraft at Salem,* first published in 1969, draws an accurate picture that links the symptoms of the afflicted to those exhibited by people suffering from the mental illness of hysteria. He treats the inhabitants of this historic period in a manner that is both respectful and knowledgeable and helps the reader see events and actions from a seventeenth century perspective. If modern knowledge about bacterial infection of the brain is added to his interpretation of the afflictions, the validity of the symptoms is reinforced [203]

**Figure 22.** *Illustration of a "fit" from J.M. Charcot, Lectures on the Diseases of the Nervous System (London, 1877). Note tongue. One study has found that one third of the people in one modern mental institution were infected with Bb.*

In 1976, Linnda Caporeal, in an article titled *"Ergotism: The Satan Loosed in Salem,"* was the first to offer an analysis of the Salem Witch Trials where the "afflicted" were treated as patients who suffered from real symptoms. Concentrating on the hallucinatory nature of some of the symptoms, her analysis included a flawed environmental study of the Salem Village area. Finding a few references to rain in the diary of Samuel Sewall, she concludes that the weather prior to the harvest of 1691 was unusually wet and conducive to the growth of ergot producing fungus on rye. This is a misreading of the evidence. Samuel Sewall also makes many notes about the persistent drought that plagued the colony. Indeed, a Palmer Summer Drought Index data set reconstructed from tree ring growth shows that Massachusetts suffered from a prolonged drought that lasted from 1680 and was only broken in late 1692. When a day of public reflection was held in Boston in 1692, it was to pray for relief from the double torments of witchcraft and drought. Notations about rain, when made, seem to underscore its scarcity and not any abundance. Even if Lyme immunosuppression is included in the picture, she fails to inventory the full list of afflicted from the records and connect them to potential sources of ergot. She does not explain why some members of a household eating from the same supply of flour remain symptom free while others become hallucinogenic. She presents what she calls a circumstantial case for the "afflicted" girls suffering from ergot ingestion. While dubious, this theory has worked its way into Salem's modern oral history of its own past. It is included in the tour script at one of the historic sites in town and was the basis for a Halloween special on television.[204]

In 1999, Laurie Winn Carlson published *A Fever in Salem: A New Interpretation of the New England Witch Trials.* She proposed Encephalitis Lethargica, caused by bites of mosquitoes as a possible source of the afflictions (again, droughts are not the best of times for explosive mosquito populations). She was the first author to suggest looking into Lyme disease.[205] Another scientific paper has suggested that the girls suffered from some form of group schizophrenia. Group hysteria has been suggested.[206] But the idea that the afflicted were actually sick has all too often been given short shrift by most modern scholarship. Mary Beth Norton, for example, states in *In the Devil's Snare* that her research into this area was limited to a conversation with a "renowned" physician who said it couldn't possibly be encephalitis![207]

But a very long list of people who were living very real lives in 1692 lay out an all too real tale of suffering. Wives lose husbands. Husbands watch their wives descend into madness and die. Children convulse and animals behave strangely. Life itself is turned upside down in some households. The root cause for some of this suffering may have been lurking in the forest edges of the Salem landscape. Many English cultural practices in the seventeenth century certainly would have been conducive to contact with ticks and exposure to Lyme disease.

Because New England is an area that is now endemic for this illness, Lyme disease should be considered as an additional diagnosis for the suffering of the humans and animals that were afflicted in the seventeenth century. While historians sometimes find solace in simplification, the real world is a complex place. Someone in 1692 could have been depressed and traumatized, have swilled down tainted rye products, and suffered from encephalitis at the same time. But there are too many similarities between the symptoms that are described in the past and those found in the very real experiences of modern people with Lyme disease to ignore.

Lyme disease is a challenge for both doctors and patients because it can be difficult to diagnose and hard to treat and can have frightening symptoms. So was suspected witchcraft. Consider James Carr, the rejected suitor who visited a physician in 1672 with a set of aggravating physical symptoms. The diagnosis that he received was that he was behagged.[208] This was a scary diagnosis that included no known medical cure and the very real threat of death. Consider Amy Tan, the modern novelist, whose early diagnosis for the cause of her Lyme disease symptoms included amyotrophic lateral sclerosis, ALS, better known as Lou Gehrig's disease. This terrifying diagnosis included little medical help and was almost certainly a death sentence.[209]

The loss of a child through miscarriage or deformity can be traumatizing, no matter what century in which it happens. Although it is an extremely rare occurrence, modern studies have documented Lyme-infected mothers and fetuses with aberrant pregnancy outcomes. There are several well described births from the Boston area and one from Maine in the seventeenth century where fetal deformities occurred which may or may not have been related to this disease. One occurred near the height of the witchcraft trials[210]

Polly Murray of Lyme, Connecticut, was part of an "afflicted" group during the last three decades of the twentieth century. She worked diligently to attract the "eye" of modern science for those who were suffering in her community. One of modern science's first answers for her problems was a stint in the Psychiatric Ward at Boston General Hospital. Seventeenth-century diagnoses like "behagged" had somehow been transformed over time into *all in your head* with the same accompanying measure of loss of control over one's life. James Carr

## AFFLICTION

probably died convinced of the fact that he had been bewitched. Polly Murray has lived to see her "afflictions" defined as Lyme disease and is an articulate and effective advocate for its modern victims.[211] Ticks, bacteria and blood have been a potent combination for centuries.

## AFFLICTION

63 See Adolph B. Benson, Ed., Peter Kalm, *Travels in North America: The English Version of 1770* (1987). Kalm reported on the fauna and flora of North America, his teacher was Carl Linnaeus.
64 Center for Disease Control, *Lyme disease* (Atlanta: U.S. Publishing Office, 2001) pamphlet.
65 W.G. Qui, D.Dykhuizen, M.Acosta, and B. Luft, "Geographic Uniformity of the Lyme Disease Spirochete and Its Shared History with Tick Vector in the Northeastern United States," *Genetics* Vol.160 (2002), 848-849.
66 Claire M. Fraser, et al. "Genome sequence of a Lyme disease spirochete, Borrelia burgdorferi," *Nature*, Vol.390 (December 11, 1997), 580.
67 W.G. Qui, op cit, "Geographic uniformity," 848-849.
68 W.G. Qiu, *Comparative population genetics of Lyme disease spirochete(Borrelia burgdorferi) and its tick vector (Ixodes scapularis) in North America*, Unpublished PhD dissertation. The State University of New York at Stony Brook.
69 A.E. Hessl and W.L. Baker, "Spruce and Fir regeneration and climate in the forest tundra ecotone of Rocky Mountain National Park, Colorado," *Arctic and Alpine Research* 29 (1997), 173-182.
70 See Map of Glaciation Maxima at 18,000 B.P on page 47.
71 T.B. Smith, R.K. Wayne, D.J. Girman, and M.W. Bruford, "A role for ecotones in generating rainforest biodiversity," *Science* 276 (1997), 1855-1857.
72 This leads to some confusion while reviewing the literature. Many early studies refer to Ixodes dammini as a separate species. It is now felt that both the southern and northern clade belong to one species, Ixodes scapularis. Qui, *Geographic*, 845-847.
73 Kathleen LoGiudice, R. Ostfeld, K. Schmidt, and F. Keesing, "The Ecology of Infectious disease: Effects of Host Diversity and Community Composition on Lyme disease Risk," *PNAS Ecology* Vol.100 No.2 (January 21, 2003), 567-571.
74 R. Ostfeld, Institute of Ecosystem Studies, Millbrook, N.Y., *Secrets of the Woods: Acorns, Biodiversity, and Lyme disease* (www.niaid.nih.gov National Institute of Health website).
75 The size of the forest is discussed in David Foster, et al. "Wildlife dynamics in a changing landscape," *Journal of Biogeography* (2002), 1339. The forest is also discussed in Tom Wessels' *Reading the Forested Landscape: A Natural History of New England* (Countryman, 2005).
76 n.a., "The effects of forest fragmentation on parasiticism of colul monkeys," The Journal of the Society of American Forestry *(*September, 2000).
77 Eric Gustafson, et al. "Evaluation of spatial models to predict vulnerability of forest birds to brood parasitism by cowbirds," *Ecological Applications* 12(2) 2002, 412-426. See also R. Ostfeld and Felicia Keesing, "Biodiversity and Disease Risk: the Case of Lyme Disease," *Conservation Biology* 14 No. 3 (June, 2000), 722-728.
78 Brian F. Allan, Felicia Keesing and Richard Ostfeld, "The Effect of Forest Fragmentation on Lyme Disease Risk," *Conservation Biology* 17, No.1 (February 2003), 267-272. See also Richard S. Ostfeld, "The Ecology of Lyme disease Risk," *American Scientist* (July-August, 1997) at www.med.harvard.edu.
79 William Bradford, *Of Plymouth Plantation* 1620-1647 (New York: The Modern Library, 1981), 73-78 and Edward Winslow, *Mourt's Relation* (Bedford, Massachusetts: Applewood Books, 1963), 18-29.
80 Daniel Gookin, unpublished papers at The Massachusetts Historical Society, *Historical Collections of the Indians of New England* (1674), 7-12.
81 Edward Winslow, Mourt's Relation (Bedford, Massachusetts: Applewood Books, 1963), 51.
82 M.L. Wilson, "Reduced abundance of adult Ixodes dammini following destruction of vegetation," *Journal of Economic Entomology* 79(1986), 693-696. If burning is not done every year, however, "Thomas Mather, David Duffy and Scott Campbell's article "An unexpected result from burning vegetation to reduce Lyme disease transmission risks" *Journal of Medical Entomology* Vol. 30 No.3 (May 1993) shows that the lower number of nymphal ticks that result tend to be mouse derived and have higher levels of infection. No one has done this study, however, for a longer span of time, so the results like everything else on this subject, are

unclear.
83 Frances Higginson wrote in 1629 that deer would be abound if it were not for the wolves, they (the deer) being in the habit of twinning or even having triplets! In Everett Emerson, ed., *Letters from Massachusetts Bay Colony 1629-1638 (*Amherst: University of Massachusetts press, 1976), 12-24.
84 Uriel Kitron, "Predicting the risk of Lyme disease: Habitat suitability," *Emerging Infectious Diseases* 8 No.3 (2002).
85 Alan Barbour and Durland Fish, "The Biological and Social Phenomenon of Lyme disease," *Science* 260 (June 11, 1993), 1610-1616.
86 Massachusetts Department of Public Health, 2002 *Lyme disease Update for Health Care Providers.*
87 A.C. Steere, E. Taylor, M.L. Wilson, J.F. Levine, A. Spielman, "Longitudinal assessment of the clinical and epidemiological features of Lyme disease in a defined population," *Journal Infect Disease* 154 (1986), 295-300.
88 David Hanley, "Vitamin D Insufficiency in North America," *Journal Nutrition February*, 135(2005), 332-337.
89 United States Department of Agriculture. *U.S. per capita Lard Consumption in pounds* 1900-2004 (2005).
90 George F.M. Ball, *Vitamins In Foods: Analysis, Bioavailability, and Stability* (UK: CRC Press, 2005).
91 FDA and EPA, Joint advisory, March 2004.
92 P.T. Liu, et al. "Toll-like Receptor Triggering of a Vitamin D-mediated Human Antimicrobial Response," *Science* 311 (March 24, 2006), 1770-3.
93 A.F. Gombart, Borregard, N., Koeffler, H.P. "Human cathelicidin antimicrobial peptide (CAMP)gene is a direct target of the vitamin D receptor and is strongly up-regulated in myeloid cells by 1,25-dihydroxyvitamin D3," *FASEB* 19, 9, (2005), 1067-77.
94 M.T. Cantorna, Hayes, C.E, and DeLuca, H.F. "Vitamin D Its role and uses in immunology," *FASEB Journal* 15, 1998(14), 25-79. Vitamin D is also a key component in the theories of Trevor Marshall, although he advises avoidance of sunlight because the cell-wall-deficient form of the Bb bacteria is capable of catalyzing the process by which Vitamin D is converted to its 1,25-D form. Instead of a slow, controlled conversion which occurs only in the kidneys, 1,25-D production becomes uncontrolled in Chronic Lyme patients, occurring throughout the body inside cells infected with cell-wall-deficient bacteria. Specifically, cells harboring cell-wall-deficient bacteria can turn into tiny, unrestrained factories producing excessive amounts of 1,25-D. Bacteria catalyze the 1,25-D conversion process intentionally to cause immune system suppression and create a more favorable living environment in the body. This theory, called the Marshall Protocol, is still subject to scientific testing and confirmation, but may be applicable to some patients.
94 J.D. Piacentino, Schwartz, B.S. "Occupational risk of Lyme disease: an epidemiological review," *Occup. Environ. Med.* 59 (2002), 75-84.
95 Ibid.
96 Ibid.
97 R. Lane, Manweiler, S., Stubbs, H., Lennette, E., Madigan, J., Lavoie, P., "Risk factors for Lyme disease in a small rural community in northern California," *Am J Epidemiol* 136(11) (Dec 1, 1992), 1358-68.
98 See David Hanley, and Davison, K.S., "Vitamin D Insufficiency in North America," *Journal Nutrition* 135 (2005), 332-337, and M. Calvo, Whiting, S.J., Barton, C.N., "Vitamin D Intake: A Global Perspective of Current Status," *Journal Nutrition* 135(2005), 310-316.
99 Karen Vanderhoof-Forschner, *Everything You Need to Know about Lyme Disease* (New York: John Wiley, 1997), 136-7.
100 Op cit., A.C. Steere, et al., "Lyme borreliosis in the Soviet Union."
101 Dr. Brian Fallon's opinions on Lyme disease are outlined in Columbia University's Alumni Magazine article *Rash judgment?* Brian Fallon '85PH, '85PS says the medical establishment underestimates the devastating effects of Lyme disease at www.columbia.edu.
102 Harper KN, Ocampo PS, Steiner BM, et al (2008). "On the origin of the treponematoses: a phylogenetic approach," *PLoS Negl Trop Dis* 2 (1): e148. doi:10.1371/journal.pntd.0000148. PMID 18235852.
103 Keynote address "The complexity of Vector-borne spirochetes" by Dr. Willy Burgdorferi given on April 9, 1999 at the Twelfth International Conference on Lyme Disease and Other Spirochetal and Tick borne Diseases, see also O. Brorson and S, Brorson, "A Rapid Method for Generating Cystic forms of Borrelia burgdorferi and their reversal," *APMIS* 106 No.12 (December, 1998),1131-1141.
104 A. Gylfe, et al., "Reactivation of Borrelia infection in birds," Nature 403 (February 17, 2000), 724-725.
105 Angela Stewart, "Study finds changes in Lyme bacteria," *The Star Ledger* (September 28, 2004), based on unpublished report from The New Jersey School of Medicine and Dentistry.
106 The full report of the Lymerix studies can be found at website of the Lyme Disease Association website: www.lymenet.org.
107 See www.sciencedaily.com/releases/2006/08/060822150046.htm

# AFFLICTION

108 Hand out- 12th Annual International Conference on Lyme disease and other Spirochetal and Tick Borne Disorders, New York, April 9, 1999. Also in Lyme disease, Survival in Adverse Conditions: The strategy of Morphological Variation in Borrelia burgdorferi and other spirochetes, 9/2003 available at www.lymeinfo.net.

[109] Isabel Diterich, Carolin Rauter, Carsten J. Kirschning, and Thomas Hartung, " *Borrelia burgdorferi*-Induced Tolerance as a Model of Persistence via Immunosuppression," *Infect Immun.* **July, 2003; 71(7):3979**–3987.See discussion of Pam3Cys [OspA]by Kathleen Dickson at www.actionlyme.org.
110 See www.AnimalNet at Scientific American Daily.
111 Tavora, F., Burke, A., Li, L., Franks, T.J., Virmani, R., "Postmortem confirmation of Lyme carditis with polymerase chain reaction," *Cardiovasc Pathol* 17 (Mar-Apr 2008), 103-7.
112 Vanderhoof-Forschner, *Everything,* 47-64.
113 Brian Fallon and J.A. Nields, "Neuro-psychiatric symptoms," *The American Journal of Psychiatry* 151, No.11 (November, 1994), 1571-83.
114 Vanderhoof-Forschner, *Everything,* 47-64.,see also Altaie,S.S., S. Mookherjee,et all. Abstract # 1-17 "Transmission of Borrelia burgorferi from experimentally infected mating pairs to offspring in a murine model.," *FDA Science Forum*, 1996.
115 Daniel Elliott, Stephen Eppes and Joel Klein, "Teratogen Update: Lyme disease," *Teratology* 64 (2001), 276-281.
116 Four out of the 13 pregnant women who participated in the Lymerix trials experienced spontaneous abortions: www.lymenet.org.
117 N.Popvic, B.Djuricic, and M. Valcic, "The Importance of Lyme borreliosis in Veterinarian Medicine," (Translated from Serbian *Glas Srp Akad Nauka 43* (November, 1993), 277-85. Karen Vanderhoof-Forschner also discusses this topic at length.
118 Mel Huff, "Symposium: Lyme disease spreading worldwide," *The Barre Montpelier, Vermont Times Argus* (August 5, 2007).
119T. J. Dunn, Jr. DVM, What is Lyme disease in dogs? Learn About this tick-borne disease, August, 2005 at www.thepetcenter.com/gen/lyme.html
120 Harald Franzen, "Ancient Tick Poses New Questions," *Scientific American* (March 28, 2001) at www.sciam.com.
121 J. Childs, *Shared Vector Zoonoses of the Old World and New World: Home Grown or Translocated: A Congressional Report* (Atlanta, Georgia: United States Center for Disease Control and Prevention, 1998), 1095-1105.
122 See Aguilar, S. 2002. "Peromyscus leucopus" (On-line), Animal Diversity Web. Accessed March 28, 2008 www.animaldiversity.ummz.umich.edu, Photograph by Calypte at www.flicker.com.
123 See British Columbia Ministry of Environment, Lands and Parks, *White Tailed Deer in British Columbia*, The Habitat Conservation Trust (2000), and also Steve Wolverton, James H. Kennedy, John D. Cornelius, "Paleo-zoological Perspective on White-Tailed Deer (Odocoileus virginianus) Population Density and Body Size in Central Texas," *Environ Manage* (2007) 39:545-552. Picture from www.animaldiversity.ummz.umich.edu
124 Dr. Kenneth Leigner lecture, April 1999 at the Twelfth International Conference on Lyme Disease and Other Spirochetal and Tick borne Diseases. The Paleolithic stretches anywhere from 750,000 to 500,000 years ago to 10,000 years B.P.
125 J. Hinnebusch and A. G. Barbour, "Linear plasmids of Borrelia burgdorferi have a telomeric structure and sequence similar to those of a eukaryotic virus," *Journal Bacteriology* 173 No 22 (1991), 7233-7239.
126 IRIN News, Reuter Foundation, *Senegal: Lyme disease: the forgotten scourge of West Africa* (August 23, 2006).
127 Christopher Lee, *1603* (New York: St. Martin's Press, 2003), 120-121.
128G. Thwaites, M.Taviner, V.Gant, "The English Sweating Sickness,"
*New England Journal of Medicine* 336(8) (Feb. 20, 1997), 580-2.
129 William Fitzhugh and Elizabeth Ward, *Vikings, The North Atlantic Saga* (Washington, D.C: Smithsonian Institute Press, 2000), 223.
130 B. Olsen, D.C. Duffy, T.G. Jaenson, A. Gylfe, J. Bonnedahl, and S. Bergstrom, "Transhemispheric exchange of Lyme disease spirochetes by sea birds," *Journal of Clinical Microbiology* 33(1995), 3270-3274.

131 See: Herbert Langer, *The Thirty Years' War* (UK: Dorset Press, 1990).
132 Norse for Deer Island
133 Nicholas Summerton, "Lyme disease in the 18th Century," *British Medical Journal* 311 (December 2, 1995), 1478.
134 See Peter Youngson, *Jura: Island of Deer* (U.K.: Birlinn Publishers, 2002).
135 B.M. Rothschild, K.R. Turner and M.A. Deluca, "Symmetrical erosive peripheral polyarthritis in the late Archaic Period of Alabama," Science 241 (1988), 1498-1501. see also B.M. Rothschild, "Tennessee Origins of Rheumatoid Arthritis", McClung Museum Research Notes, No.5, April 1991. Tennessee is part of the original home territory of the Cherokee Native Nation.
136 B.A. Lewis, "Prehistoric juvenile rheumatoid arthritis in a pre-contact Louisiana native population reconsidered," American Journal Physical Anthropology 106 (June 1998), 229-248, Lewis hypothesizes that the spirochete Borrelia burgdorferi was the infectious agent responsible for prevalence of adult rheumatoid arthritis in prehistoric southeastern Native American populations, 2) that Bbis a possible cause of the arthritis evident in individual16ST1-14883b, and 3) that antibodies to B. burgdorferi provided partial immunity to the related spirochete Treponema pallidum (syphilis) for the precontact Tchefuncte population from Louisiana, protecting them from severe treponemal response. Given the probable widespread existence of Ixodid tick vectors for B. burgdorferi in prehistoric North America, coupled with the existence of syphilis, it follows that the transition of Native American hunting-gathering economies to more sedentary economies would predictably be linked to an increased incidence of syphilis due to the loss of benefits of the above-stated partial immunity. Inferences regarding biological controls interacting with and influencing prehistoric Native American migration patterns are suggested from the link of B. burgdorferi to an Ixodid tick common to northeast Asia.
137Columbus wrote this in a letter to the King and Queen of Spain on November 19, 1502 during his fourth (and final) voyage to the New World. (See Warren White's *La Vizcaina: Investigative report based on evidence and hypothesis* at www.geocities/warren white for a description of the trip). This letter may have prevented the royal couple's endorsement of any fifth voyage.
138 A witch panic occurred in Bermuda in 1651. See Behringer, *Witches and Witch-hunts (*Malden, Mass.: Polity, 2005), 145.
139See J. Terborgh, "The role of ecotones in the distribution of Andean birds," *Ecology* 66 (1985), 1237-1246, also see maps of migratory routes, Two other important studies are K. Kurtenbach, et al., "The Key roles of selection and migration in the ecology of Lyme borreliosis," *International Journal Medical Microbiology* 291 Sup.33 (June, 2002),152-4 and J. D. Scott, "Birds Dispurse Ixodid and Borrelia burgdorferi infected ticks in Canada," *Journal Med. Entomology* 38 No.4 (July, 2001), 493-500. The stress of migration may activate Bb in birds. See S. Millus, "Migration may reawaken Lyme disease," *Science News* (February 19, 2000).
140 Jonas Waldenström, Åke Lundkvist, Kerstin I. Falk, Ulf Garpmo, Sven Bergström, Gunnel Lindegren, Anders Sjöstedt, Hans Mejlon, Thord Fransson, Paul D. Haemig, and Björn Olsen, "Migrating Birds and Tickborne Encephalitis Virus," *Emerging Infectious Diseases* Volume 13, Number 8 (August 2007). See also D, Richter, A. Spielman, N. Komar, F.R. Matuschka, "Competence of American robins as reservoir hosts for Lyme disease spirochetes," *Emerging Infectious Diseases*, Vol 6, Number 2 (March-April 2000),133-8.
141 Carville Earle, "Environment, Disease and Mortality in Early Virginia" in *The Chesapeake in the Seventeenth Century* (Chapel Hill, N.C.: University of North Carolina Press, 1979), 96-125. He argues that the early settlers suffered from salt poisoning via their drinking water. He also discusses the high level of food and protein intake during Jamestown's 'starving times.'
142 Edward Winslow writes about his experiences during a trip to visit Massasoit's Village in *Mourt's Relation* (Bedford, Massachusetts: Applewood Books, 1963), 67.
143Brown and Bowen in Ronald L. Michael, ed., "Animal Bones from the Cross Street back lot privy," *Historical Archeology* 32 (3) (1998), 77.
144 Paul Boyer and Stephen Nissenbaum, eds., *Salem-Village Witchcraft: A Documentary Record of Local Conflict in Colonial New England* (Boston: Northeastern University Press, 1993), 132-133.
145 Ticks, when noticed, were sometimes called wood lice in the 17th C. See Adolph B. Benson, Ed., Peter Kalm, *Travels in North America*.
146 See *Revolutionary War Journal of John Ford*, unpublished manuscript in the National Park Service collection, Morristown, N.J. for a description of his encounters with ticks; also see National Archives, *Revolutionary War Pension Applications*.
147 Robert Remini, *The Life of Andrew Jackson* (New York: Penguin, 1990), 7-9. See also A.S. Kesselheim, *Deception and Presidential Disability: A Historical Analysis* (Trans Stud College Physicians: Philadelphia, Dec 23, 2001), 87-98.
148 Loge, Ronald V. "Illness at Three Forks: Captain William Clark and the First Recorded Case of Colorado

# AFFLICTION

Tick Fever," *Montana: The Magazine of Western History,* Vol. 50, #2 (Summer 2000), 2-15. William Clark lived until 1838. Although there is no mention of Meriwether Lewis suffering from tick fever, he suffered from some form of mental illness or dementia and committed suicide on October 9, 1809. Neuroborreliosis should be added to bi-polar disorder and malaria as a potential diagnosis for Lewis' illness.
149 A.B.R. Myers, *On the etiology and prevalence of diseases of the heart among soldiers* (London, John Churchill, 1870), 4.
150 See National Archives, *Civil War Veteran Census* (1890). Also, see unpublished *Civil War Diary of Egbert Oswald Hixon, March 27-August 11, 1862* for a description of encounters with ticks during the Peninsular Campaign at www.nextech.de/ma15mvi/biogrphy/hixoneo.htm.
151 W.F. Marshall III, S.R. Telford III, P.N. Rhys, B.J. Rutledge, and D. Mathiesen, "Detection of Borrelia burgdorferi DNA in museum specimens of Peromyscus leucopus," *Journal of Infectious Disease* 170 No.4 (Oct. 1994), 1027-32.
152 F.R. Matuschka, A. Ohlenbusch, H. Eiffert, D. Richter, and A. Spielman, "Characteristics of Lyme disease spirochetes in archived European ticks," *Journal of Infectious Disease* 174 No.2 (August 1996), 424-6.
153 Bruce Rothschild,"The Tennessee Origins of Rheumatoid Arthritis,"*McClung Museum Research Notes,* No.5, April 1991.
154 Shepard Kretch III, *The Ecological Indian, Myth and Reality* (New York: Norton, 2000), 164-70.
155 Michael Weiner and Janet Weiner, *Herbs that Heal* (Quantum Books: Mill Valley, California, 1994), 228,264,270,278.
156 CDC website at www.cdc.gov.
157 *See Daryl Hall Cancels concerts due to Lyme disease* (7/7/2005) at www.hallaandoates.com.
158 Tad Reeve, "Herron, down with Lyme disease, has picked himself up just in time," *Saint Paul Pioneer Press,* MN. (August 10, 2004).
159 Miranda Hitti, "Lyme Disease Benches FSU Football Quarterback" in *Experts Discuss Lyme disease Symptoms,* Web MD Medical News (July 11, 2005).
160 Dawn Irons, "Tennessee Governor Hospitalized with Illness from a Suspected Tick Bite" (August 17, 2006) on *Fox News.*
161 Polly Murray, *The Widening Circle* (St. Martins Press: New York, 1996).
162 Amy Tan quoted in J.J. McCoy,"Lyme diagnoses can miss the bull's eye ", Maine Sunday Telegram(August 17,2003), 1G.
163 Dr. Fallon's work is listed at www.columbia-lyme.org. Preliminary findings of a SPEC brain scan study of Lyme patients done at Columbia University and Columbia Presbyterian Hospital in New York City has been completed and is awaiting publication at the time of the writing of this book.
164 Pictures of ECM's can be seen online at www.lymediseaseassociation.org/PhotoAlbum_Rash.html
165 Arno Karlen, *The Biography of a Germ* (New York: Anchor, 2001), 133.
166 See Christina Larner, *Enemies of God,* (Edinburgh: Scotland, UK: John Donald, 2000), 11.
167 Olaus Magnus, *Historia de Gentibus septentrionalibus* (1555).
168 Lambert Deneau, *A Dialogue of witches(*1575).
169 Ludovico Maria Sinistrari, "Demoniality" in Masters, R.E.L. *Eros and Evil: The Sexual Psychopathology of Witchcraft* (AMS Press, Reprint edition. 1995).
170 Reginald Scot, *Discovery of Witchcraft (*1584).
171 See Cotton Mather, letter of May. 31, 1692 to John Richards, in Silverman, *The Life and Times of Cotton Mather (*New York: Columbia University Press, 1985), 37.
172 Del Rio, *Disquisitionum Magicarum* (1599).
173 Brad Paisley song *"Ticks"* lyrics, see www.bradpaisley.com.
174 See Christina Larner, *Witchcraft and Religion: The Politics of Popular Belief* (UK: Blackwell, 1986).
175 Dr. Jacques Fontaine, *Des Marques des sorciers et de la réele possession que le diable prend sur le corps des homes* (1611).
176 Peter Ostermann, *Commentarius Juridicus* (1629).
177 Margaret Alice Murray wrote *The God of the Witches* and *Witch Cult in Western Europe.* Her works have been routinely ridiculed but she did make the extremely important contribution to the subject of witchcraft in that she was the first to demonstrate that the Christian concept of the devil's physical appearance including having horns, a tail, cloven feet, and holding a pitchfork has roots in pagan fertility gods.
178 The rash on the child's cheek can be compared with descriptions of the "tawny red spot" on the cheek of a dying child that helped set off a witch-hunt in Connecticut in 1662.
179 R.R. Müllegger, "Clinical aspects and diagnosis of erythema migrans and borrelial lymphocytoma," *ACTA 10*, No 4 (2001).
180 See list in Appendix G.

181 See 1662 in Appendix G.
182 Mary Hortado in Burr, *Narratives,36*
183 Jarvis Ring, "Testimony," Essex County, Massachusetts Archives, *Salem Witchcraft* Vol 1: No. 181.
184 Deodot Lawson quoted in Burr, Narratives, 53.
185 Noel Ivor Hume, *A Guide to the Artifacts of Colonial America* (Philadelphia: University of Pennsylvania Press, 1969), 254-56.
186 This is from personal observation of attached ticks on family members and myself. They can look like little dark brown splinters or sticker thorns.
187 Emerson Baker, *The Devil of Great Island* (New York: Palgrave, 2007), 189-190.
188 Cotton Mather, *A Brand plucked out of the burning*, 264.
189 Full text of Robert Calef, "More Wonders" is in George Lincoln Burr's *Narratives of the New England Witchcraft Cases (*Mineola, New York: Dover, 2002).
190 The Shattuck child, Burr, *Narratives*, Ibid, 225-226, 380.
191 Ibid, *Narratives*, 157.
192 A swollen gland in a throat feels like a little egg. See Essex County Archives *Salem Witchcraft* Vol. 1, 63-64.
193 Ibid, *Narratives*, 287.
194 See Nikhil V. Dhurandhar, "Infectobesity: Obesity of Infectious Origin," *Journal of Nutrition* 131(2001), 2794S-2797S. Infectobesity is a concept that is also being researched by Dr. Ritchie Shoemaker.
195 Denise Lang, *Coping with Lyme disease* (New York: Holt, 2004), 138.
196 List is included in chart, *Compendium Maleficarum*.
197 List of symptoms from Karen *Vanderhoof-Forschner, Everything You Need to Know about Lyme Disease and Other Tick Borne Disorders* (New York: John Wiley & Sons, 1997), 47-64.
198 Denise Lang, *Coping with Lyme disease* (New York: Holt, 2004), 122.
199 Many of the extant witchcraft trial papers are in Parris' handwriting. This is evident when you when compare the writing with sermon books that Parris wrote during the same time period.
200 John Demos, *Entertaining Satan: Witchcraft and the Culture of Early New England* (New York: Oxford Press, 1983).
201 Paul Boyer, and Stephen Nissenbaum, *Salem Possessed: the Social Origins of Witchcraft* (Cambridge, Massachusetts: Harvard University Press, 1974).
202 Mary Beth Norton, *In the Devils Snare* (New York: Vintage, 2003).
203 Chadwick Hansen, *Witchcraft at Salem* (New York: George Braziller, 1969).
204 Linnda Caporael, "Ergotism: The Satan Loosed in Salem?,"
*Science* 192 (1976).This theory has a life of its own, refuted in Nicholas Spanos and Jack Gottlieb's "Ergotism and the Salem Witch trials" in *Science 194*, it then rose again in Mary Matossian's *Poisons of the Past: Molds, Epidemics and History* (New Haven: Yale University Press, 1989).
205 Laurie Winn Carlson, *A Fever in Salem* (Chicago: Ivan R. Dee, 1999).
206 Thurman Sawyer and George Bundren, "Witchcraft, Religious Fanaticism and Schizophrenia—Salem Revisited," *The Early American Review* 3 No. 2 (Fall, 2000), Article 5.
207 Norton, *Snare*, Introduction
208 James Carr testimony, Essex County Archives, *Salem Witch Trials* Vol. II, 38.
209 Amy Tan quoted in J.J. McCoy, "Lyme diagnoses can miss 'bulls-eye': Author Amy Tan finds herself among Unconventional Cases," Maine Sunday Telegram (August 17, 2003), 1G.
210 John Winthrop wrote a gossipy detailed account of Mary Dyer's stillborn daughter in his journal and in several letters: "It was a woman child, stillborn, about two months before the just time, having life a few hours before...it was of ordinary bigness; it had a face, but no head, and the ears stood upon the shoulders and were like an ape's; it had no forehead, but over the eyes four horns, hard and sharp, two of them were above one inch long, the other two shorter; the eyes standing out, and the mouth also; the nose hooked upward all over the breast and back, full of sharp pricks and scales, like a thornback; the navel and all the belly with the distinction of the sex, were where the back should be; and the back and hips before, where the belly should have been; behind, between the shoulders, it had two mouths, and in each of them a piece of red flesh sticking out; it had arms and legs as other children; but, instead of toes, it had on each foot three claws, like a young fowl, with sharp talons." Mary Dyer lived outside Boston at Romney Marsh. Cotton Mather's wife also delivered a deformed child on March 28, 1693, that died within a matter of days. While no elaborate description survives, Samuel Sewall stated that the child had no anus. Cotton Mather had multiple contacts with "afflicted" persons during the 1680's and 90's, including bringing some "afflicted" children to live in his home. Another monstrous birth was recorded by Rev. Moody as having occurred in Berwick, Maine in Mather, *Illustrious Provinces,* 167 and *Collections of the Massachusetts Historical Society* (1968), 361-62.

# AFFLICTION

211 See Polly Murray, The *Widening Circle*, for a vivid description of her experiences, 18-30.

## II. PREDICTING RISK: GEOGRAPHY, MASTING, MOISTURE, AND LIVING ON THE EDGE

Adult *Ixodes* ticks are brought into geographic areas primarily by three hosts: birds (which can also carry immature stages of ticks into distant areas, especially during migrations), mice, and deer- that are capable of ranging over wide areas, especially along riparian corridors. To become successfully established in any area, however, ticks must have a suitable habitat for all of their life stages- questing, molting, over wintering, and egg laying. Vegetation, soil, topography and climate are strongly related to habitat suitability for these ticks.[212] The range of *Ixodes* ticks seems to be expanding. They have recently been found in some areas, such as northern Sweden,[213] that were once too cold for ticks to survive in, a finding that may suggest both a modern change in climate and an

# PREDICTING RISK

expanded or more dispersed range during the warm climate that occurred during the medieval time period.

Several factors have been identified by modern science as being crucial for the maintenance of a robust tick population and the resulting witches brew of human risk for Lyme disease. Tick densities have been found to be highest in areas which are associated with soil textures of increased particle size: sandy soils. The most in-depth modern study of habitat suitability for *Ixodes scapularis* done to date is from Wisconsin. It found the highest level of ticks in deciduous forests. Ticks were nearly absent in clay soils and most abundant in sandy loam textured soil. Although they are found in all quaternary deposits, glacial moraine sites were most heavily populated by these ticks.[214]

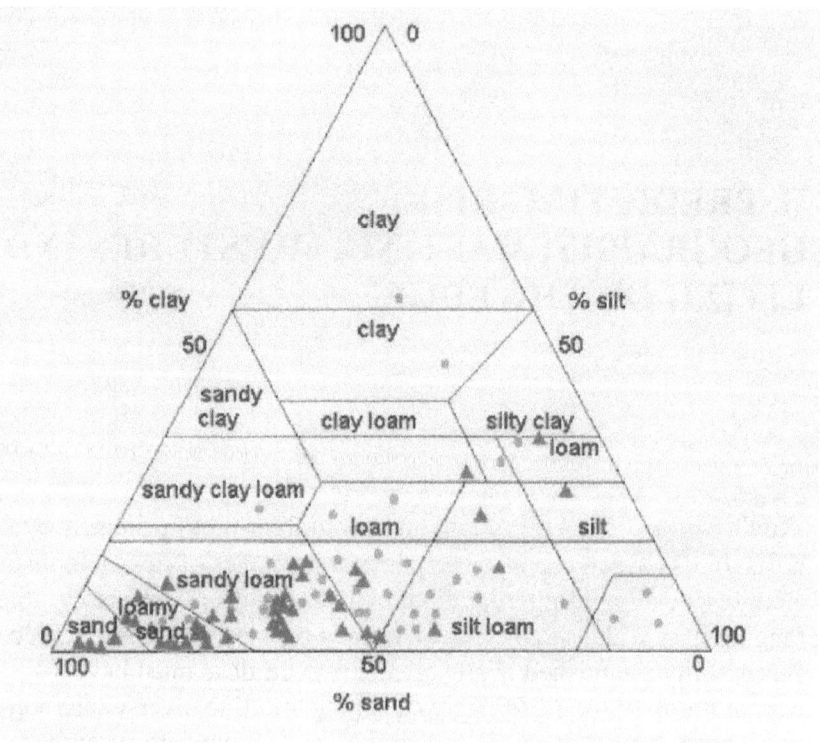

**Figure 23**. *Soil Texture study from Wisconsin shows that Ixodes scapularis prefer sandy loamy soils. A triangle indicates sites with ticks present, a circle denotes site with no ticks present. Emerging Infectious Diseases Vol. 8, No. 3, March 2002.*

 DISGUISED AS THE DEVIL

This information corresponds favorably with information from both modern and historic Massachusetts. The entire state was covered by a glacier during the last Ice Age. Cape Cod and the Plymouth area have a Pleistocene and Holocene sedimentary base and soil that is dominated by the sandy, well drained Carver category. It takes thousands of years under normal erosive condition for significant changes to occur in the soil itself. The modern soil map, except for areas of urban disturbance, can be comparable to the soil map from the seventeenth century. The soil that covers Cape Cod, foe example, is estimated be between 5,000 and 15,000 years old.[215] The Salem and Salem Village area has a more mixed bedrock underlay with a variety of sandy loam soils. Modern hot spots for Lyme disease in Massachusetts include Cape Cod and the Ipswich and Haverhill areas that are near Salem.[216]

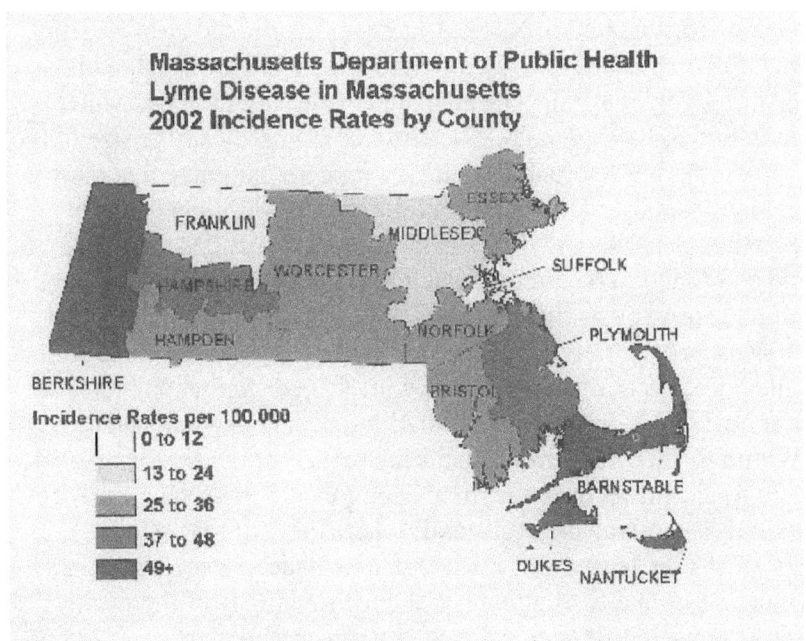

**MAP 3**: *Massachusetts Department of Public Health 2002.*

Because leaf litter is a necessary component for the survival of over-wintering deer ticks, tick densities tend to be highest in forests that are dominated by deciduous trees and lowest in coniferous forests.[217] Oak (*Quercus*) forest ecosystems dominated the landscape of seventeenth

## PREDICTING RISK

century Massachusetts and today cover much of the eastern United States. Oak forests create a swath of forest that girds the earth to include parts of North America, Europe, and Asia. The interactions between oak forests and other landscape components has also been the subject of modern research. Despite the complexity of these systems, these studies have revealed that strong interactions among a relatively small set of species have profound consequences for Lyme disease prevalence.

Oak trees are important for a variety of reasons. In addition to providing leaf litter, they episodically produce large acorn crops called "mast" every two to five years, which sets off an ecological chain reaction. Acorns are rich in fat, are large when compared with other seeds, and, as any squirrel knows, store well over the winter due to delayed germination caused by the presence of tannins. Acorns are a highly preferred food for many forest species.

The population response in mice to abundant mast production has been well documented in North America, Europe, and New Zealand. The population density of the white-footed mouse in New England is affected strongly by masting. The annual peak density of mice, which is achieved in mid to late summer, is determined primarily by the size of the acorn crop during the prior autumn. In one modern experiment, a mast year was simulated by adding acorns under mature oaks in a designated area. The winter, spring, and early summer mouse densities were then measured and found to be three to five times higher in the experimental area than on control areas that had received no acorn supplementation.

Mast production also causes a strong behavioral response in white-tailed deer. In normal conditions white-tailed deer tend to have an average annual home range from one-half to one and one half square miles. Within this territory, movements are related primarily to seasonal changes in food sources or cover. In the autumn of a mast year, deer are attracted to oak-dominated forest stands where they forage on the abundant acorns. In autumns of low acorn abundance, deer spend less time in oak forests and their autumn diet consists largely of the woody browse that grows in tree stands that are dominated by maples.

Because of these effects, masting strongly influences both the dynamics of *Ixodes scapularis* ticks and the risk of exposure to Lyme disease. Adult *Ixodes* ticks are deer specialists. They quest in the autumn by climbing up shrubby undergrowth and waiting for a warm-blooded vertebrate, which is usually a deer, to brush by. Once onboard a deer, the adult female tick anchors her mouth parts into the skin and takes a blood meal that lasts about three to five days, during which males, who have

## DISGUISED AS THE DEVIL

also jumped on but feed only intermittently, will mate with the attached female. The male ticks soon die. The female tick drops off the deer and over-winters in leaf litter on the forest floor in a quiescent, engorged state.

The following spring, the female will lay a mass of up to a few thousand eggs before she dies. These eggs hatch in midsummer into the larval stage. They are usually born in an infection free state. They immediately commence host seeking at ground level close to where they hatched. White footed mice, who also occupy the forest floor at this same time, are highly likely to become a larval tick's first blood meal host. Approximately forty to eighty percent of larval ticks feeding on an infected white-footed mouse will acquire the *Bb* spirochete during their first blood meal. Because of this high infectivity, the white footed mouse is considered to be the principal natural reservoir for Lyme disease in the Eastern and Central United States.

After a two to three day blood meal, larval ticks drop off, and molt several weeks later into the nymphal stage that over-winters in the forest floor. The following spring and early summer, these nymphs become active and again seek a blood meal host. Nymphal ticks are not terribly selective, but because they quest at ground level, the hosts they are most likely to encounter are small mammals and ground-dwelling birds. If the nymphal tick acquired the Lyme spirochete during its larval meal, it will retain that infection throughout its life and may also infect its host during the nymphal blood meal. Most human cases of Lyme disease are traceable to bites by these infected nymphs. After a three to four day blood meal, the nymphal tick drops off, molts into the adult stage, and begins questing again that same autumn.

Space use by deer determines where in the landscape this will happen. The location of deer in the fall, when adult ticks are active, determines where in the landscape eggs will be laid and larval ticks will hatch the next year. When deer occupy oak-forest patches in the autumn of a mast year the result is a heavy concentration of larval ticks in oak forests the following summer. An experimental simulation of a mast year resulted in larval tick burdens on mice that were forty percent higher on acorn-addition plots compared to controls. [218]

The spatial pattern of a forest edge landscape often takes the form of discrete patchiness that is also important to animal population dynamics. In some parts of the Salem area in 1692, oak dominated woodlots existed in an agricultural and developing matrix, whereas in other zones, small, managed agricultural fields existed in a forest matrix.

## PREDICTING RISK

In both cases, patches of oak forest often were juxtaposed with patches of other forest types, abandoned Native American sites, clear cuts, and the like. Many of these patch types must have provided suitable habitat for both white-footed mice and white-tailed deer. The mobility of mice and their tendency to disperse during phases of rapid population growth mean that the ecological dynamics within these oak patches are connected to those of adjacent patch types in the landscape. Masting not only affects mouse density within oak-dominated forest stands, but also appears to affect the later density of mice in neighboring patches, which in turn spreads the risk for human contact with infected ticks. [219]

Other studies have looked at the effect of moisture on modern Lyme disease incidence. The reported Lyme disease incidence between 1993 and 2001 in seven northeastern states was analyzed as an outcome of weather variability. For all seven states analyzed, including Massachusetts, relationships were found for the correlation of early summer disease incidence with the June moisture index or Palmer Hydrological Drought Index in the region 2 years previously. When rainfall levels were compared with Lyme disease incidence rates for several Eastern states, results have been found to be are statistically significant for three states: New York: $r2 =.74, p=0.003$, Conn.: $r2 =.45, p=0.046$, and, Mass.: $r2 =.49, p=0.036$. The relationship was not significant for three other states that were studied, Rhode Island, Maryland, and Pennsylvania. These correlations seem to reflect enhanced nymphal tick survival in moister conditions under some circumstances.[220] Moisture affects not only tick survival but also spatial distribution. In drought situations, mice may leave a territory totally[221] but deer tend to forage in wider areas and move around more. This, in turn, creates a wider area where lower numbers of infected ticks may occur and will include the same water source riparian areas that may also be used by humans. In some states, significant correlations were also observed related to warmer winter weather a year and a half prior to disease incidence, which may have been due to higher survival and activity levels of the white-footed mouse, the main host for the *Bb* spirochete.[222]

Can this complex web of information be applied to landscape of the past? Bog core pollen samples from Cape Cod show that it was covered by an oak/ scrub pine forest in 1620.[223] This information fits the written descriptions of both Bradford and Winslow. Forest edged environments are described, as well as a Native American altered landscape consisting of forest and open patches. Winslow wrote that the

 DISGUISED AS THE DEVIL

harbor (near modern Provincetown) was encompassed about to the very sea with oaks, pines, junipers, sassafras, and other sweet wood. The wood was "for the most part open and without under wood, fit either to go or ride in." Deer were found near a pond where "many vines and fowl and deer haunted there." At one part of an initial exploratory journey, Bradford noted, "they fell into such thickets as were ready to tear their clothes and armor in pieces." Cape Cod had some of the highest modern rates for Lyme infection in 2002. (See Map 3)

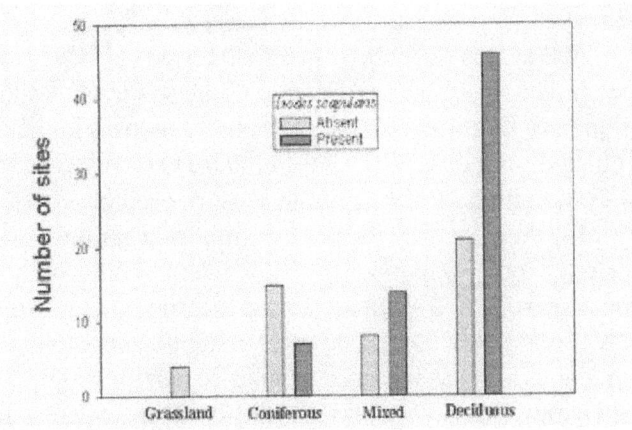

**Figure 24.** *Land Cover categories and number of sites with Ixodes scapularis present and absent. Ticks favor deciduous forests. From "Predicting the Risk of Lyme Disease: Habitat Suitability for Ixodes scapularis in the North Central United States" in Emerging Infectious Diseases Vol. 8, No. 3, March 2002.*

Pollen analyses from archeological work done during the 'Big Dig' at the site of a late seventeenth century dwelling's privy vault in Boston, tested exploratory core pollen percentages for between about 1650 and about 1690. It shows a forest dominated by an oak/pine mix with an interesting absence of pine pollen in the upper most level. Most of the pines in the area may have been harvested for naval wares by 1692, leaving an almost totally deciduous forest for a period dominated by oaks with very few maple trees.[224]

One of the most useful tools for modern Lyme disease epidemiologists has been the Geographic Information System (hereafter GIS). Using a Landsat Thematic Mapper, various elements can be factored into satellite geological system generated risk maps to predict the presence of the optimum tick vectors.[225] This type of analysis is slated to be undertaken in New Hampshire in the future because the incidence of Lyme disease is on the rise in that state.

This has been done very effectively in Westchester County, New York, a Lyme disease endemic area. Information from the ground, developed through drag sampling and canine serology reports was compared and correlated with the satellite generated GIS map data for risk. In a study of 337 residential properties, high-risk properties consistently tested as being both significantly greener than low risk properties. High-risk properties appeared to contain a greater proportion of broadleaf trees, while low risk properties had more non-vegetative cover and open lawn. [226]

The model that was developed was then tested further by comparing it with data about human infection that was obtained through a random questionnaire that was sent out to property owners in the area. By comparing predicted with observed (questionnaire responses) data, a 71 percent accuracy was found for the remote sensory prediction of risk.[227] Another study, also done in Westchester County, used canine seroprevalence rates to analyze the effect of residential adjacency to forest on Lyme disease risk. This study found that the rate of Lyme infected dogs was positively correlated with living adjacent to the forest.[228]

A study done in Maryland used ground-based zip codes to map annual risk for Lyme disease. It found that the greatest risk areas for actual infection occurred along green riparian features, especially along the watershed areas for Gunpowder River and Deer Creek near Baltimore. This was well correlated with a GIS assessment for the area. These riparian zones were regularly used by deer and other wildlife as a source of drinking water.[229]

Another study done in Maryland concluded that contact between humans, ticks, and wildlife vector hosts is facilitated in landscapes with high forest- herbaceous interspersion or edge environments. Both deer and white footed mice thrive in these heterogeneous landscapes. Deer prefer a mix of forest cover and open areas with tender vegetation. The white footed mouse is highly opportunistic and will inhabit both forest edges and open patches. All of these habitats may, by coincidence, be adjacent to human habitation areas and Lyme disease is also strongly associated with peri-domestic exposure. This spatial pattern is most likely to occur in a modern suburban setting and historically in a dispersed settlement pattern. While both masting and moisture levels play important roles in Lyme disease infection rates, land cover patterns are also significant features for this infection. In one study, Lyme disease rates were found to

 DISGUISED AS THE DEVIL

be highest where there was the greatest amount of edge interspersion between herbaceous (lawn) and forest with a ratio that included forty to sixty percent herbaceous cover. The analyzed spatial characteristics plot as being high risk or low risk.[230]

In the modern landscape, a particular set of biological and environmental circumstances has converged to create the perfect conditions for a Lyme disease outbreak. Forest has returned to land that was cleared in the 18th and 19th century for farming and dispersed suburban neighborhoods have developed. There is now more forest in many places in New England than at any time in the previous one hundred and fifty years.[231] This secondary forest occurs patchily amid occupied spaces creating an edge habitat,[232] which is ideal for deer. The deer population has exploded because they have no remaining natural predators in New England and hunting is restricted in suburban areas. In modern times the tick population has expanded along with that of the deer, and Americans became exposed to Lyme disease again in the late twentieth century when they moved en masse into the deer/tick habitat of the suburbs. A similar story occurred in the growing suburbs of Europe.

Figure 25.

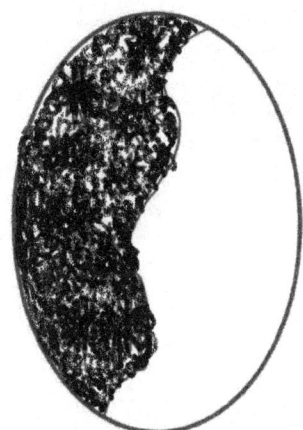
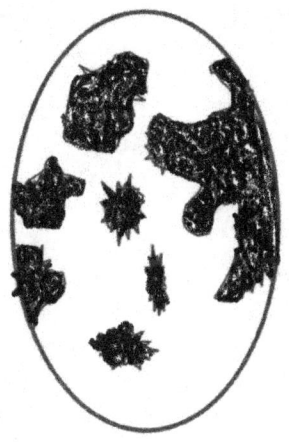

LOW RISK
Low forest-herbaceous edge
50% forest/50% grass/lawn
Low forest fragmentation

HIGH RISK
High forest-herbaceous edge
50% forest/50% grass/lawn
High forest fragmentation

## PREDICTING RISK

The placement of roadways onto a landscape may also alter the distribution, abundance and behavior of wildlife species that are involved in Lyme disease. Compared to the surrounding landscape, roadways often constitute a different environment. This is an area that is also in need of much further research.[233]

The modern map of Lyme disease risk in the United States is heavily focused in the New England and Mid Atlantic area. This can be compared to a historic map of affliction and witchcraft accusations for the same area, which shows a close geographical correlation. Almost all of the historic witchcraft accusations in the English colonies of North America occurred in areas of high modern predicted Lyme disease transmission risk.[234] Only the accusations from Virginia and North Carolina lie outside of this high-risk zone.

**MAP 4: GEOGRAPHY OF LYME DISEASE:** *Lyme Disease Risk Map-U.S.A, Highest risk area (in black) includes parts of New England.*

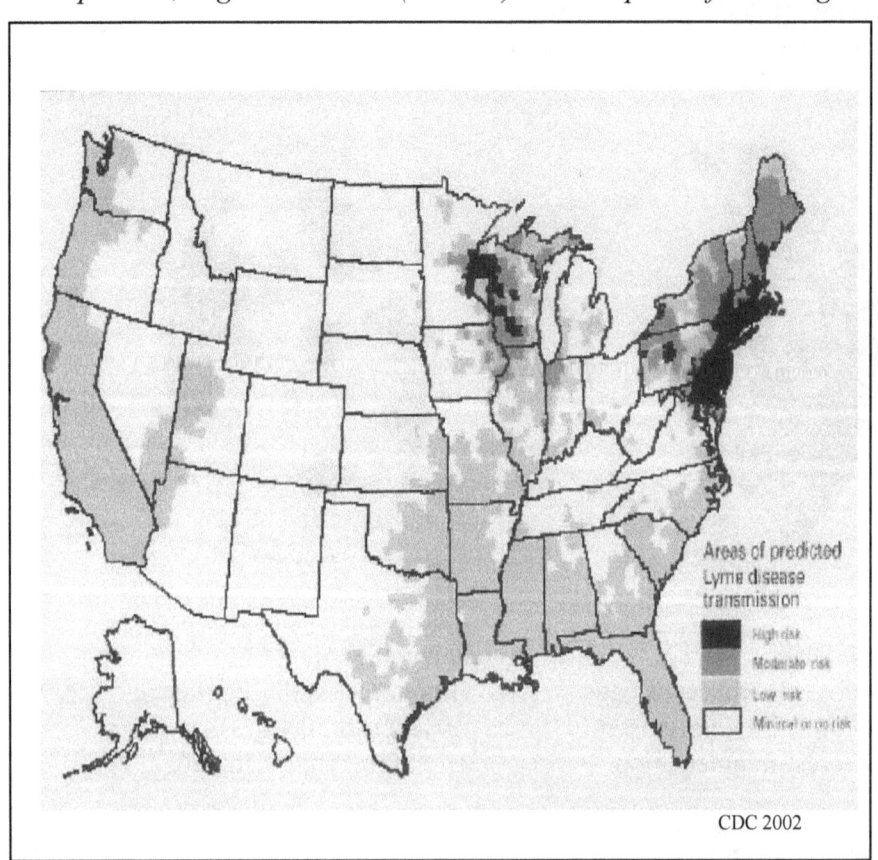

MAP 5: **GEOGRAPHY OF WITCHCRAFT:** *The Algonquin Linguistic Native American Group that the Creek and Cherokee belonged to inhabited what is now the entire Eastern United States, from the Great Lakes down towards the Gulf Coast, prior to the arrival of European settlers. Witchcraft accusations [in **black**] in the English Colonies of North America were concentrated in New England.*

## PREDICTING RISK

In Europe, the highest rates of borrelia infection of *Ixodes Ricinus* ticks are found in countries in central Europe, especially in Germany, with smaller hot spots in Sweden, Norway, Estonia and coastal Portugal. *B. burgdorferi* is found in Europe but *B. afzelii* and *B. garinii* are the most common *Borrelia* species, but the distribution of genospecies seems to vary in different regions in Europe. In a recent study that surveyed scientific literature, the infection rate was higher in adult ticks, with

**MAP 6: GEOGRAPHY OF LYME DISEASE:** *Map of defined regions of infected ticks in Europe. Areas with lower infection rates are indicated by smaller letter size, areas with higher infection rates are indicated in gray with larger sized letters. Spain and Southern Italy have a negligible infection rate. SEE APPENDIX E for list.*

 DISGUISED AS THE DEVIL

an average of 18.6 percent, than in nymphs, at 10.1 percent. No difference between collection period (1986 to 1993 versus 1994 to 2002) was found.[235] This can be compared with a map of the intensity of historic witch hunting in Europe and a similar geographic correlation is found, as in North America.

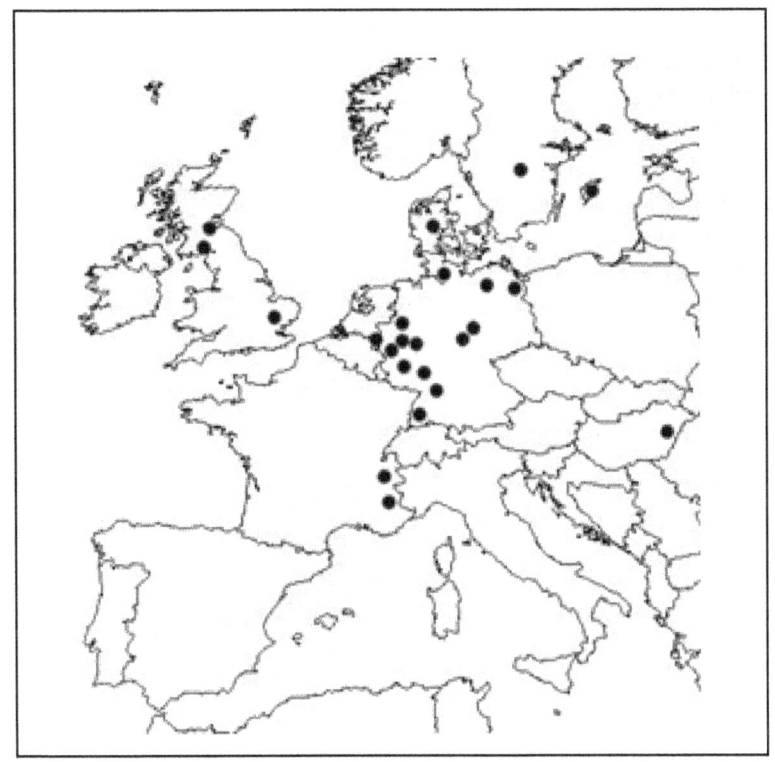

**MAP 7: GEOGRAPHY OF WITCHCRAFT *IN EUROPE DURING THE LITTLE ICE AGE*:** *Hot Spots of witch-hunts in black, 1400-1800 A.D.*
NWDIII

Comparing the average infection rate for adult and nymphal ticks from this modern meta-analysis with the estimated ratio rates for witchcraft executions per number of inhabitants set at the 1600 date, but arranged according to modern geographic boundaries, results in a statistically significant relationship being found between the two variables. The places with higher levels of witchcraft persecution in historic Europe are also places associated with higher levels of infected ticks in the modern landscape.[236]

## PREDICTING RISK

This association may underlie one of the puzzles related to European settlement in the New World. Both Spanish and French settlements had very low levels of witch accusation activity, especially in contrast with the English colonies in New England. This might be expected of the Spanish, who did little witch hunting in Spain itself but the French certainly participated in witch-hunts and helped write the books about witches. The French, however, largely settled in what was [in the cold Little Ice Age post 1492 time period] poor tick habitat- either too far north (Quebec area), too far south (New Orleans), or were exploring too far west (The Great Plains) to encounter many infected *Ixodes* ticks. They never came down with 'Devil's Marks,' had pins stuck into their skin, had neurological problems or any of the other symptoms associated with witchcraft afflictions. Spanish colonists, once settled in, were simply too far south to experience many afflictions and any accusations that were made, in places like New Mexico, seem to center

***Map 8. Territories Owned by Spain between 1516 and 1556:***
*In addition to land in Spain, areas that are shaded in on this map were also part of the Kingdom of Charles V. They include places with the highest number of witch prosecutions recorded in Europe and also correspond to areas with high levels of infected Ixodes ticks today. The virulent Osp C clone type A form of Borrelia burgdorferi has been found in modern France, Austria, Germany, and Northern Italy.*[237] NWDIII

## DISGUISED AS THE DEVIL

around the suppression of Native shamanism, not disease. When Spanish territory is located in good *Ixodes* tick habitat in northern latitudes, however, there is an opposite effect. The level of witch-hunting skyrockets.

These areas that were part of the Spanish Kingdom of King Charles V, who ruled from 1516 to 1556, for example, were key witch hunting zones, especially after Columbian contact with the New World. Interactions with the islands in the Caribbean that are located along bird migration routes may have introduced ticks and a virulent form of Lyme disease into the Rhine River Valley after 1492. During the sixteenth century Antwerp became the world's great center of commerce, finance, and contact with the New World. The definition of the European witch, including symptoms and the presence of the devil's mark, would largely come to be written in these same areas.

Obviously, a satellite-generated map of the seventeenth-century Massachusetts landscape is not available. However, the University of Virginia is in the process of creating a GIS examination of Salem Village and the surrounding area. Although certain features, most importantly waterways, have surely altered position in the past 300 years, there are some historic features from 1692 that have survived to the present day. These include five houses, the known site of the Parris parsonage foundation, Wills Hill, Humphreys Pond Island, and Wenham Lake. By aligning these features with their placement on the historic Upham map, an analytical comparison was made.

This analysis found that there were some inaccuracies in the placement of the known still extant feature points on Upham's map, ranging from 50.387 meters off mark for the Edward Bishop House to 221.724 meters for Will's Hill. It found that comparative distance measures on Upham's map were most accurate in the east and less accurate in the west. However, the relative shape distribution of the points on the map and the GIS control map matched fairly well: there were no dramatic surprises.[238]

With this in mind, working with the maps that are available, but realizing their flaws, and then adding in descriptions about landscape features written by people on the ground in their historic time, using pollen studies for a determination of forest composition, and soil surveys for a determination of underlying soil composition, it is possible to generally map areas of approximate high risk for Lyme disease infection- a geography of risk for the past. This results in a risky map for western Salem Village.

## PREDICTING RISK

The recreated map for late 17th century Salem reveals a mosaiced landscape with forest fragmentation, edge environments, and riparian corridors.[239] In 2001 the Essex County area of Massachusetts had a reported average infection rate of twenty five to thirty six cases of Lyme disease per population of 100,000 with a probable actual incidence rate of 250 to 360 per 100,000 or 2.5 to 3.6 per thousand based on CDC estimates of underreporting, Plymouth chalks in at 3.7 to 4.8 per thousand and Cape Cod is at 4.9 or more.[240]

Moisture and temperature can also be examined. Wolfgang Behringer correlates a relationship between these factors and witch hunting in *Climatic Change and Witch-hunting: the Impact of the Little Ice Age on Mentalities*. In Europe during the late 14th and 15th centuries, as we shall see later, the traditional conception of witchcraft was transformed into the idea of a great conspiracy of witches partially in response to "unnatural" climatic phenomena. Weather making was listed among the traditional powers attributed to witches and demonology underlay any analysis of severe weather in the medieval European mindset. Extended witch-hunts took place at various peaks of the Little Ice Age because a part of society held witches directly responsible for climatic anomalies and their impacts, especially disease prevalence. [241]The body of evidence shows that a large region of mid and northern Europe experienced a sharp cooling at around 1570/80 that, at least in the north, marked a shift towards a prolonged period of cool weather conditions. This region had its southern boundary in the Alps and there is little evidence for either a major cooling in southern Europe or much interest in witch-hunting.[242]

The documentary data used for the reconstruction of seasonal and annual precipitation and temperatures in central Europe (Germany, Switzerland and the Czech Republic) during the age of the great witch-hunts include narrative sources, several types of proxy data and 32 weather diaries. Results were compared with long-term composite tree ring series and tested statistically by cross correlating with documentary data based on instrumental measurements taken in the twentieth century. It was shown that the series of indices could be taken as good substitutes for instrumental measurements. Results show that in the sixteenth century winter temperatures remained below the 1901-1960 average except in the 1520s and 1550s.

Summer climate was divided into three periods of almost equal length. The first was characterized by fluctuations between cool and warmer seasons. The second interval was warmer and between five and six

 DISGUISED AS THE DEVIL

percent drier than in the 1901–1960 period. Summers after 1560 were colder and four percent more humid. Autumns were colder in the 1510s and twenty percent wetter in the 1570s. Attempts to compare circulation patterns in the sixteenth century with twentieth-century analogues revealed that despite broad agreements in pressure patterns, winters with distinct northeasterly patterns were more frequent in the sixteenth century, whereas the declining summer temperatures from the mid-1560s seem to be associated with a decreasing frequency of anti-cyclonic ridging from the Azores' towards continental Europe. The number of severe storms on the Dutch North Sea Coast was four times greater in the second half of the century than in the first. A more or less continuous increase in the number of floods over the entire century occurred in Germany.

The analysis of the effects of climate on rye prices in four German towns involved a model that included monthly temperatures and precipitation values known to affect grain production. The correlation with rye prices was found significant for the entire century. Rye reached its highest values between 1565 and 1600. From the 1580s to the turn of the seventeenth century wine production slumped almost simultaneously in four regions from Lake Zurich to western Hungary. This also had far-reaching economic effects and caused a temporary shift in drinking habits from wine to beer. Stressed peasant communities that suffered collective damage from the effects of climatic change pressed authorities for the organization of witch-hunts. It was during this time period that witch affliction was codified into a list of the symptoms that are included in works like the *Compendium Maleficarum of 1608* [See list at the end of this chapter].[243]

Temperature and moisture records can also be extrapolated from tree ring data for New England. The growth of various species of trees is effected in measurable ways that can be translated into Palmer Drought Index moisture level and temperature measurements. The summer drought index data for a tree in Worchester, Massachusetts is included in Appendix E. Measurements for Cape Cod are also available. These measurements show that in Massachusetts the temperatures from 1675 through 1697 were cold with 1677 and 1692 being the coldest years. This is consistent for the Little Ice Age period. Massachusetts also suffered a drought with lower than average rainfall for a prolonged time period starting in 1681 that only ended in the summer of 1692.[244] This prolonged drought is mentioned in the diaries of both Cotton Mather and Samuel Sewell and must have had a profound effect on local food production.

## PREDICTING RISK

When the Massachusetts Bay Colony called for a day of prayer and fasting in 1692, it was for the double plagues of Witchcraft and drought.[245] This witch drought is particularly obvious in the tree ring records.

Using the regression formula included in a study that relates Lyme disease incidence purely to moisture levels for modern Massachusetts:[246]

*Lyme disease Incidence= 500.393+ (126.55) times the (June Palmer Drought Index of two years earlier) per 100,000 population,*

a crude comparable infection rate can be computed for the seventeenth century using PHDI levels obtained by examining tree rings from 2 years prior. For 1691, based on 1689 ring measurements, the Worchester area would have seen about four cases per 1000 population and 4.5 cases per thousand would have occurred on Cape Cod, for an average of slightly over four cases per thousand residents that year for the Massachusetts Colony. The population of Salem Village is estimated to be between 550 and 650 people[247] so it would be a bit over 2 cases for that year.

More importantly, the witch drought would have created clusters and concentrations of ticks along waterways that were used by thirsty deer, mice, and humans. Masting levels also affected where ticks would have been concentrated. In Salem Village, this could have occurred near the Ipswich River on the western, splotchily developed, side of the village, increasing the risk of Lyme disease there above the four per thousand in population that is predicted.

# PREDICTING RISK

212 Marta Guerre, Uriel Kitron, et al., "Predicting the Risk of Lyme disease: Habitat Suitability for Ixodes scapularis in the North Central United States," *Emerging Infectious Diseases* 8, No.3 (March 2002): see chart on page 114.
213 Elisabeth Rosenthal, "Global Warming: Adapting to a New Reality," *The International Herald Tribune* (September 15, 2005).
214 Guerre, et al, *Predicting Risk*.
215 See *Geologic Map of Massachusetts*, State of Massachusetts Public Information Department, also *Soil Survey of Barnstable County and Plymouth County in Massachusetts*, United States Department of Agriculture Soil Conservation Service (September, 1986).
216 See *Soil Survey of Essex County*, Massachusetts, the Southern Part, 1984, Sheet 7.
217 Ostfeld, Richard. "The Ecology of Lyme disease Risk," American Scientist (July-August, 30. 1997).
218 Ibid.
219 Ostfeld, Richard, Brian F. Allan, Felicia Keesing, "The Effect of Forest Fragmentation on Lyme Disease Risk," *Conservation Biology* 17, No.1, (February, 2003).
220 Susan Subak "Effects of Climate on Variability in Lyme Disease Incidence in the Northeastern United States," *Am J Epidemiol* (2003), 157:531-538.
221 Jones, C.J. and U.D. Kitron, "Populations of Ixodes scapularis are modulated by drought at a Lyme disease focus in Illinois," *J. Med. Entomol.* 37(2000), 408-415.
222 Subak, *Effects of Climate*.
223 Glenn Motzkin and David Foster use pollen core samples to recreate the historic landscape of Cape Cod in "Grasslands, heathlands and shrublands in coastal New England: historical interpretations and approaches to conservation," *Journal of Biogeography* 29 (2000), 1569-1590.
224 Gerald Kelso, "Pollen Analysis of the feature 4 privy at the Cross Street Back Lot Site, Boston, Massachusetts," *Historical Archeology* 32, No. 3, (1998).
225 M.C. Nicholson and T.N. Mather, "Methods for Evaluating Lyme disease risks using GIS and geospatial analysis," *Journal Medical Entomology* 33 No. 5 (September, 1996), 711-720.
226 Louisa Beck, Bradley Lobitz and Byron Wood, "Remote Sensing and Human Health: New Sensors and New Opportunities," *Emerging Infectious Diseases* 6 No. 3 (April 2000) at www.cdc.gov.
227 S.W. Dister, D. Fish, S.M. Bros, D.H. Frank and B.L. Wood, "Landscape characterization of peridomestic risk for Lyme disease using satellite imagery," *American Journal of Trop. Med. Hyg.* 57 No. 6 (December 1997): 687-92.
228 Louisa Beck, *CHAART Lyme disease New York- Municipality Study of Lyme disease* (July, 2000) at http://geo.arc.nasa.gov linked canine seropositivity with peridomestic risk for Lyme disease.
229 Christina Frank, Alan Fix, Cesar Pena, G. Thomas Strickland, "Mapping Lyme Disease Incidence for Diagnostic and Preventive Decisions, Maryland," Emerging Infectious Diseases 8 No. 4 (April 2002) at www.cdc.gov.
230 Laura E. Jackson, *The relationship of Land Cover patterns to Lyme disease* unpublished PHD dissertation, UNC at Chapel Hill, (2005).
231 Foster, David R. "New England Wide Regional Studies at Harvard Forest," *Harvard Forest* 2005 (April 25, 2005). At http://harvardforest.fas.harvard.edu/research/newengland.html
232 See Laura Jackson, Elizabeth Hilborn, James Thomas, *The Relationship between Lyme disease and Landscape Development Pattern*, EPA science forum poster handout.
233 P.D. Haemig, Waldenstrom J, Olsen B "Roadside ecology and epidemiology of tick-borne diseases." *Scand J Infect Dis*, Jul 11, 2008.1-6.
234 The CDC figures tick infection rates as well as prevalence of established populations as factors that lead to predicted Lyme disease transmission risk levels.
235 Caroline Rauter and Thomas Hartung, "Prevalence of Borrelia burgdorferi Sensu Lato Genospecies in Ixodes ricinus Ticks in Europe: a meta-analysis," *Applied and Environmental Microbiology*, Vol. 71, No. 11 (November 2005), 7203-7216.

236 At a ninety five percent confidence level, the r square is .192, p=.047. See Wolfgang Behringer, *Witches and Witch Hunts, A Global History* ( Malden, Mass.: Polity, 2004) chart on page 150, discussion of how numbers were determined 149-160.

237 W.G. Qui, J.F. Bruno, W. McCaig,Y.Xu,I.Livy, M. Schriefer, and B. Luft, "Wide Distribution of a High Virulence *Borrelia burgdorferi* Clone in Europe and North America,"*Emerging Infectious Disases,* Vol. 14, No.7, July 2008,1097-1104.

238 The University of Virginia Salem Witchcraft GIS project information can be found at http://lewis.lib.virginia.edu.

239 Boyer and Nissenbaum, *Salem-Village Witchcraft*, 130.

240 CDC, *Lyme disease Incidence Map*, Massachusetts Department of Public Health.

241 Wolfgang Behringer, "Climatic Change and Witch-hunting: the Impact of the Little Ice Age on Mentalities," *Climatic Change* Volume 43, Number 1 (September 1999), 335-351.

242 K.R. Briffa, P.D. Jones, R.B. Vogel, F.H. Schweingruber, M.G.L. Baillie, S.G. Shiyatov, E.A. Vaganov." European Tree Rings and Climate in the 16th Century," *Climatic Change* Volume 43, Number 1 (September 1999), 151-168.

243 Christian Pfister, Rudolf Brázdil. "Climatic Variability in Sixteenth-Century Europe and its Social Dimension: A Synthesis," *Climatic Change* Volume 43, Number 1 (September 1999), 5-53

244 See Briffa et al. European tree rings, 2929-2941.

245 Both Cotton Mather and Samuel Sewell write about the day of fasting on March 31, 1691/92 for the double problems of drought and witchcraft.

246 Subak, Susan "Effects of Climate Variability in Lyme disease Incidence in the Northeastern United States," *American Journal of Epidemiology* Vol.157, Number 6(2003), 531-538, G. E. Glass, et al., "Environmental risk factors for Lyme disease identified with Geographic Information Systems," *American Journal of Public Health* 85 No. 7 (July 1995), 944-948,and R. Ostfeld, Secrets of the Woods: Acorns, Biodiversity, and Lyme disease, NIAID archives at www.niaid.nih.gov. see also Wolff J. O. "Co-existence of white-footed mice and deer mice may be mediated by fluctuating environmental conditions," *Oecologia* 108(1996), 529–33.

247 Based on Upham's map and records in Boyer and Nissenbaum's *Salem-Village Witchcraft: A Documentary Record of Local Conflict in Colonial New England*. Boston: Northeastern University Press, 1993.

# DISGUISED AS THE DEVIL

| EVIL POSSESSED/ BEWITCHED[248] | LYME DISEASE |
|---|---|
| It does not, like natural sicknesses, come on by degrees; but the sick man often suffers the severest symptoms from the very beginning | Symptoms can appear suddenly and be debilitating |
| The patient's sickness is very difficult to diagnose, so that the physicians hesitate and are in doubt and keep changing their minds, and are afraid to make any definite pronouncement about it | Poor diagnostic tests, can be difficult to diagnose, physicians disagree on treatment |
| The sickness is very erratic: and although it may be periodic, it does not keep its regular periods | |
| Witches squint or evil eye, prefer darkness to daylight | Symptoms can wax and wane |
| | Photophobia |

## DISGUISED AS THE DEVIL

Often depicted as crippled, by the 1400's witches are shown flying through the air on poles, brooms, and on the backs of animals

Bewitched cont'd

Witches have quarrelsome personality

The possessed, being but ignorant, argue about high and difficult questions; or when they discover hidden or long forgotten matters,
or future events, or the secretes of the inner
conscience, such as the sins and imaginings of
the bystanders; or if they provoke then to
quarrel without cause or become so furious
that they cannot be bound or restrained by many strong men, they hear a voice speaking inside them

Lyme cont'd

Personality changes, mental changes

# A COMPARISON

Arthritic Lyme can limit ability
to walk

Bewitched cont'd

Witches eat babies, cause
infertility and hideous
deformities in live births:
in our own time many children
have been
born with two heads, with six
fingers, with
two bodies, and with other limbs
duplicated
in a marvelous manner or else,
they are
lacking in the necessary and
usual equipment
of the human body[249]

Another deformity described:
that shapeless
mass like a palpitating sponge or
marine
zoophyte with every evidence of
life unknown to them[250]

Lyme cont'd

Spontaneous abortion can occur
and birth defects have been
noted

## DISGUISED AS THE DEVIL

up their food. Some are afflicted
with violent fever
and headache, and their whole
body is weakened
and in pain, the sick man speaks
in foreign tongues

Bewitched cont'd

Lyme cont'd

If something moves about the
body like a
Live thing, so that the possessed
feel as it
were ants crawling under their
skins….
Palpitating like a fish….If the
part of the
body…is stirred by a sort of
Palpitation…If the patient is
tortured with
certain prickings

Peripheral neuropathies

The sicknesses with which those
who are
bewitched suffer are generally a
wasting         or
emaciation of the whole body
and a loss
of strength, together with a deep
languor, dullness of mind,
various melancholic ravings,
different kinds of fever, such a
weakness pervades the whole
body that they can hardly move
on any
account at all, lose their appetite,
some vomit

Fatigue, fever, flu like
symptoms, brain fog, mental
confusion

# A COMPARISON

Bewitched cont'd | Lyme cont'd
---|---
It is as though wind descended from his head to his Feet, and then again went from his feet to his head… If he feels as if cold water were continually being poured down his back | Chills
In some the throat is so constricted that they seem as if they are being strangled, some feel a rising and falling in their throat | Sore throat, Cranial Nerve Palsies
Certain convulsive movements of an epileptic appearance, a sort of rigidity of the limbs giving the appearance of a fit | Seizures
Sometimes the head swells in all directions when his brain feels as if it were tightly bound, or pierced and stricken as if by a sword. | Encephalitis, meningitis

## DISGUISED AS THE DEVIL

Some inner power seems to urge
the possessed to hurl himself
from a precipice, or hand or
strangle himself or the like

Bewitched cont'd

Sometimes the whole skin, but generally
only the face, becomes yellow or
ashen colored

Lyme cont'd

Liver involvement-jaundice

Sometimes they become as if
they are stupid,
blind, lame, deaf, dumb, lunatic,
and almost
incapable of movement.
If the demon takes hold of his
tongue and twists
it and makes it swell or if the
mouth is stretched
wide open and the tongue thrust
out

Cranial Nerve Palsies, joint pain,
Multiple system involvement

# A COMPARISON

Depression/
Suicide

Bewitched cont'd

Some have their hearts punctures as if by
Needles, some feel as if their heart was being eaten away

Many feel acute pain in their guts, from the Abdominal orifice of some there issue
certain matters like balls, as if they were worms or ants or frogs

Stolen penises

Lyme cont'd

Heart Involvement, heartburn

Irritable Bowel Syndrome, digestive problems, constipation

Erectile dysfunction

---

[248] The symptoms listed for the past are from Guazzo, *The Compendium Maleficarum,* 1608, a list of 'How to distinguish Demoniacs and those who are simply bewitched', Book III, Chapter 2, 167-170, and descriptions of witches included in the Salem Witch Trial testimonies, see Appendix G, 251-70. These were based upon the collective knowledge acquired over several hundred years of experience defining witchcraft and demonic possession. The symptoms for Lyme

disease are from Karen Vanderhoof-Forschner, *Everything You Need to Know about Lyme Disease* (New York: John Wiley, 1997) and a list derived from treatment guidelines developed by Dr. Burrascano, 11/2002, see Appendix J.

[249] This describes the outcome one of (the Quaker) Mary Dyer's pregnancies. See Endnote 200.

[250] This describes the 1638 outcome of one of accused witch Ann Hutchinson's pregnancies.

DISGUISED AS THE DEVIL
# THINKING IN TIME

Lyme disease is an affliction that is linked with the seasons. Trying to figure out a time frame for affliction in the distant past is complicated by the calendar change of 1752. During the Middle Ages, church officials and scholars noted that church holidays did not occur in their appropriate seasons (think Christmas in July). In 1582, Pope Gregory XIII and his astronomer and mathematician created a reformed calendar known as the Gregorian or New Style (N.S.) calendar. It was adopted first in Roman Catholic countries but not in Protestant countries until the eighteenth century. England and its American colonies adopted the reformed Gregorian calendar in 1752. To make this calendar adjustment, eleven days were dropped from the month of September in 1752 and a leap day was added every four years from that point on in time. The start of the year was moved from March 25 to January 1. So the 31st of December, 1751 was followed by the start of a new year on the 1st of January 1752 and the 2nd of September, 1752, was followed by the 14th of September, 1752. Double dating was used in Great Britain, colonial British America, and British possessions to clarify dates that occurred between 1 January and 24 March in the years between 1582 and 1752, including 1692. To correlate past seasonality with the modern seasons it is important to factor in the added days plus a leap day every four years up to 1752. Everything may have been a bit colder during The Little Ice Age. The Jamestown landing of May 14, 1607 in Virginia was seasonally earlier in modern April. The Pilgrim landing on November 11, 1620 is roughly equivalent to the end of September-not quite so close to being on the brink of winter as our modern mythology would suggest. The Puritans began to land at Massachusetts Bay in the modern seasonal equivalent of late April and Betty Parris began to show symptoms of affliction in the modern seasonal equivalent of late autumn, 1691.[251] These are all times of the year when nymphal or nymphal and adult *Ixodes* ticks can be actively questing for blood meal hosts.

---

[251] Duncan, David Ewing. *Calendar: Humanity's Epic Struggle to Determine a True and Accurate Year* (New York: Bard, 1998).See also Smith, Mark M. "Culture, Commerce, and Calendar Reform in America," *William and Mary Quarterly* 3rd Vol. 55 (October, 1998), 558-84.

ENGLISH SETTLEMENT

## III. ENGLISH SETTLEMENT IN THE NEW WORLD:

Pease porridge hot, pease porridge cold,
Pease porridge in a pot nine days old.
Some like it hot, some like it cold,
Some like it in a pot nine days old.[252]

                      Old English Nursery Rhyme

The story of the founding of the English colonies in the New World during the first half of the seventeenth century is one of the many 'histories' that make up a larger mythic American identity. Many parts of this foundation myth have been analyzed, questioned, embellished, interpreted and re-interpreted for over three hundred years. Two parts of the story that have been passed down in a relatively intact manner are the 'facts' that the first settlers at Plymouth and in the Massachusetts Bay area were subject to periods of starvation and that the initial great sickness in both colonies was caused by scurvy. Food plays a key role in this mythology.

     This chapter will analyze both the deeply rooted cultural subjectiveness of seventeenth-century English food preferences and the actual diet that was being eaten in this early time period, as described by the settlers themselves. Since scurvy is now known to be caused by a deficiency of vitamin C in the diet, any possible sources of this nutrient will be examined. It will also propose Lyme disease, contracted after contact with a 'geography of risk' for this ailment, as an alternate diagnosis for the root cause of the initial period of illness in both Plymouth and Massachusetts Bay and as the cause of sporadic outbreaks of illness throughout the seventeenth century, and as a possible cause of the afflictions of the deponents in the Salem Witchcraft Trials of 1692. It follows a trail of evidence that begins with food.

     Exactly what is considered to be food can be a highly subjective, culturally determined concept. The first settlers to come to Massachusetts brought with them a set of cultural attitudes related to both food and foodways. The traditional diet of the seventeenth-century middle class yeoman in England was based largely on legumes and grains. Studies of records from medieval archaeological sites show a high level of cereal in the diet.[253] A typical meal was porridge or pottage, a thick soupy stew made up of a grain (usually oats) or legume base with some onion or other vegetables mixed in along with the occasional chunk of expensive beef or pork, with a hunk of coarse bread on the side. Beer was brewed from barley and hops. Add in some dairy products made from cow, goat, or sheep's milk and a meal was complete. Bread was endorsed in biblical terms as the "staff of life." Christ himself beseeched God to "give us this day our daily bread." This diet created a grain-centric cultural mindset that was carried within each English settler who stepped foot ashore in North America.[254]

Shipboard food mimicked the land diet as closely as it could but was definitely high in salty preserved meat and lacking in the fresh fruits and vegetables that would have been available on land. The lack of vitamin C in this diet made for a prevalence of shipboard scurvy, especially during long journeys. It takes three months of vitamin C deprivation for scurvy to set in. Scurvy is a disease that is characterized in later stages by mouth sores and hemorrhages (mucosa hemorrhage), loose teeth, extreme fatigue and prostration, and ecchymoses--a form of macula (spot, blemish or stain) appearing in large irregular shaped hemorrhagic areas of the skin. It is preceded by a period of ill health characterized by sallow complexion, loss of energy, and pains in the extremities and joints.[255] Scurvy is one of the oldest diseases known to humankind. There is evidence of this affliction in the Old Testament, the Ebero Papyrus, and the writings of Pliny. Scurvy is considered to have been endemic in Northern Europe during the late winter months in the Middle Ages. The first concise account of scurvy was written in the 13th century by Jacques de Vitry's in the *History of the Crusades*. He wrote:

> *A large number of men in our army were attacked also by a certain pestilence, against which the doctors could not find any remedy in their art. A sudden pain seized the feet and legs; immediately afterwards the gums and teeth were attacked by a sort of gangrene, and the patients could not eat any more. Then the bones of the legs became horribly black, and so, after the greatest patience, a large number of Christians went to rest on the bosom of the Lord.*[256]

Scurvy haunted ships crossing between England and early Massachusetts and was a definite concern for travelers. The first outbreak of sea scurvy had been recorded during Vasco da Gama's expedition to India in 1497,

## ENGLISH SETTLEMENT

which lasted four months and had a mortality of 93 deaths out of 148 crewmembers, chiefly from scurvy.[257] It has been argued that scurvy, which killed two million sailors between 1500 and 1800, should be classified as history's foremost occupational disease. The Dutch were the first to recognize the antiscorbutic effect of oranges and lemons. Berries were prized as being antiscorbutic in Northern Europe, as were rose hips, which are a rich source of vitamin C.[258] By 1620, scurvy was considered to be somehow connected to both diet and sea travel. Several remedies for this side effect to cross-Atlantic travel were included in the prevailing English body of common knowledge. These included lemon juice, and teas made from lemon grass, sassafras, and rose hips. All are sources of some vitamin C except the watered lemon grass.[259]

After a fairly healthy two month (scurvy usually takes three months to set in) journey across the Atlantic, both Edward Winslow and William Bradford state that the occupants of the *Mayflower* began to become ill soon after they landed at Cape Cod. Winslow states that they suffered from "coughs and colds, the weather proving suddenly cold and stormy, which afterwards turned to the scurvy."[260] Bradford wrote that half of his company died "being infected with the scurvy and other diseases which this long voyage and their inaccommodate condition had brought upon them." Bradford connects the illness with sea travel, poor living conditions and a lack of proper food and drink (beer). More than one third of the early chapters of Bradford's *Of Plymouth Plantation* revolve around attempts to acquire or grow food, especially grain.[261]

In Europe at any time during the Middle Ages the primary determining factor for diet was class. The laws instituted in England after the Norman Conquest in 1066 reserved the meat of wild mammals like boar, rabbits and deer for the aristocracy and designated "parks" and certain forests as exclusive hunting grounds. Because of this, higher classes ate higher percentages of meat than lower classes. One study of account records from 1338 shows meat and fish making up fifty eight percent and bread twenty six percent of the diet of an upper class household, while meat made up twelve percent and bread and grains sixty four percent of the diet for the lower working class. If these percentages held through time, the middle class seventeenth century English settlers would have been accustomed to eating a diet with a similar high percentage of grain-based foods.[262]

Once on shore in New England, the Pilgrims immediately identified and accepted maize as an edible grain, renaming it Indian corn after the common corn (wheat) of Europe and the Native Americans whose caches

## DISGUISED AS THE DEVIL

they had looted. The rest of their initial diet revolved around this grain, allowing the settlers to retain an acceptable English grain-centric food model with little alteration. Over time, they substituted a new world bean, calling it a pea bean, for the 'pease' in their porridge, to create the now classic New England Baked Beans. Again, an indigenous food substitute was renamed to occupy an already accepted role in the colonists' diet.[263] Any understanding of the concept of having a poor diet and hard or starving times, as described by the first English settlers, needs to be examined with this cultural bias and mindset included in the interpretation. Starving times can more accurately be described as 'lacking grain times.' The English would go to great lengths to obtain grain-based food. Thirteen years earlier at the settlement in Jamestown, Virginia, a lack of grain made those desperate Englishmen vulnerable to "sailors who would pilfer biscuits to sell, give, or exchange (with the settlers) for money, furs, sassafras, or love."[264] In Plymouth, William Bradford, greeting a ship with seven new mouths to feed, bemoaned the lack of anything to make bread. However, at the same time (Spring 1622), the same ship, the *Sparrow,* laden with a cargo of fish but "no victuals nor any hope of any," was allowed to leave Plymouth to sell the fish in Virginia. Apparently fish were considered to be barely fit for an English stomach.[265] (Chemical analysis of the composition of ancient human bones from 183 early Britons show that this preference for grains [and land animals] over seafood began, even in the coastal areas of England, over 6,000 years ago[266] and was probably well engrained in the English psyche by the seventeenth century.)

At another point in time, Bradford sent Edward Winslow to Monhegan Island, off the coast of Maine, to procure what provisions he could from the English ships that were fishing in the area. He returned with a store of meal that they could spare (but not fish).[267] The Pilgrims traded with the Native Americans for corn. When a Native woman offered two traveling Pilgrims a platter of corn and dried oysters, they accepted the corn but gave away the oysters to their Native traveling companions.[268]

The English considered fish a "difficult" food because it was hard to cook, since it tended to fall apart and was full of bones. And then it had to be eaten. This was difficult when the only utensils available were knives, spoons and fingers. Fine as a trading commodity, its proper place was on someone else's plate. The food of penance in pre-reformation England, fish continued to carry this stigma into the seventeenth century. It was reluctantly eaten, replaced by other foods when they became more available, and would not reappear with any frequency until the early nineteenth century.[269] This is clearly delineated to have happened in the Massachusetts Bay settlement. Archaeological research done at a seventeenth-century Boston privy (Cross Street 4) shows that fish accounted for 6.9 percent of the total consumed bio-mass (food from animal sources) in the earliest levels (+/- 1650) but dwindled erratically down to 0.6 percent by the dawn of the eighteenth century.[270]

Little did William Bradford know that the boost in human vitamin D levels caused by the inclusion of so much seafood in the initial diet might have contributed to the newfound health of Plymouth's fledgling population. The almost homogenous middle class status of those first settlers in Massachusetts (in contrast to Jamestown's fifty percent population of upper class "gentlemen" settlers)[271] made their initial eating of fish and 'wild' meats a marked departure from the normal diet they had eaten in England. A normal diet would be once again assumed when more grain and domesticated animals became available in the New World.

The hunting of deer, for example, was not only illegal for the lower classes but was seen as the exclusive sport of a decadent upper leisure class in England. Because of their class status, the settlers had not participated in this pastime in England. Moreover, because the Pilgrims and Puritans had not been deer hunters back in England, these first settlers were probably without much cultural interest in hunting them in the New World.[272] Although John Winthrop's first meal in the New World was "a good venyson pastye,"[273] he probably considered any consumption of wild flesh to be a temporary stopgap measure needed to fill the void caused by a shortage of domesticated cattle. This shortage was expressed in 1630 when Winthrop wrote to his wife "we are here in paradise, though we have not beef or mutton."[274]

In the archaeological work done in association with Boston's 'Big Dig', the food remains in several seventeenth-century privies were examined. They show that venison was an extremely rare item in the prevailing foodways of Boston between 1650 and 1700. Out of hundreds of faunal remains found, deer bones turn up only three times, including once in the Katherine Nanny privy and again in a privy belonging to a wealthy pewterer.[275]

This lack of interest in shooting deer was clearly shown by the actions of the set of armed Englishmen who made one of the first exploratory trips on Cape Cod in 1620. They were properly armed (with fowling pieces and a spaniel bird-hunting dog) to hunt (and did eat) fowl during their trek. When they saw three deer, Edward Winslow wrote that he would like to have taken one of them, but no one actually fired a shot. Perhaps

## DISGUISED AS THE DEVIL

their muskets were too heavy and awkward to quickly aim and fire. Maybe their fowling pieces had shot that was too light to kill a deer. Another explanation might be that they had little inclination to shoot because they were simply not culturally molded to be deer hunters. They were not technically on a food procurement foray. However, when they found a stored cache of Indian corn nothing kept them from taking it. To a grain-centric mind, the deer with its wild meat could easily be ignored, but the corn was much too tempting to leave behind.[276]

Hard times in Plymouth Plantation came whenever grain was in short supply or gone. There was, however, a seemingly adequate supply of other, less desirable, foodstuffs that actually created a nutritionally adequate diet. The one small boat that Plymouth had, while not "overwell fitted," was sent out net fishing for bass and other fish. If the boat was late coming back or had a poor catch, "all went seeking shellfish," which they "digged out of the sands" at low tide. In the winter they would gather groundnuts (Jerusalem artichokes) and, of necessity, shoot fowl and deer, although they appear to have obtained most of their venison from their Native neighbors.[277]

On the arrival of the ship *Anne* in 1623, Bradford stated, "for food they [the settlers at Plymouth] were all alike, save some that had got a few pease of the ship that was last here. The best dish we could present our friends with was a lobster or a piece of fish without bread or anything else but a cup of spring water. And the long continuance of this diet, and their labors abroad, had abated the freshness of their former complexion; but God gave them health and strength in good measure, and showed them the truth of that word that man liveth not by bread only." He continued, "God fed them out of the sea for the most part."[278] In other words, in their "sad state" they were tanned, healthy and strong, but had by necessity, and not choice, a diet that depended upon seafood of all types and lacked bread. Even worse, there was no beer! To a grain-centric mind, this was as close as it could get to starving. Considering the "sad state" Bradford felt their diet was in, it is easy to understand why he felt that they were suffering from scurvy! (Although skin manifestations are never mentioned anywhere in the historic records for this contact period, a comparison between photographs of the skin hemorrhages that are a symptom of scurvy and the ECM of Lyme disease shows great similarities in appearance).

Looking at this diet with twenty first century eyes, it appears to be an early high protein- low carbohydrate diet that might be admired by the likes of the late Dr. Atkins, washed down with lots of pure free spring water (we now buy it by the bottle). Transition to this diet would have initially been accompanied by the very real physical phenomenon know as "carbohydrate craving," caused by a drop in blood insulin levels (the induction phase in the Atkins' diet), but even those pangs would diminish over time as the body became accustomed to a lowered insulin response. [279] After the great sickness of the first winter, and except for normal fevers and such, the residents of Plymouth seemed to remain remarkably healthy and many lived long lives. This may have been one of the most direct results of their diet.

Bradford himself questioned the depth of Jacob's Biblical famine. "They had such great herds and store of cattle of sundry kinds, which, besides flesh, must needs produce other food as milk, butter and cheese, etc. and yet it was counted a sore affliction." His poor colony of Plymouth "not only wanted the staff of bread but all these things [meat and dairy products]" and were stuck eating seafood. Like any other English settler, Bradford must have longed for the homey comfort of a bit of beef and a slab of bread slathered with a thick slice of English cheese washed down with a nice warm beer.[280]

In turn, twenty-first century eyes can question the depth of Bradford's sore afflictions. Surely if his people were starving, he, as Governor, would not have allowed even the pesky adventurer Thomas Weston to send a ship full of edible fish to Virginia when it was needed to feed people at home. In actuality, the food consumed during the first years of settlement, as described by William Bradford, Edward Winslow and others, shows a healthy nutrient level, high levels of vitamin D and more than adequate amounts of vitamin C.

The early settlement period in what is now the state of Massachusetts was marked by an initial phase of dramatically high mortality rates followed by a more normal and predictable cycle of disease and mortality. At Plymouth, after a healthy eight-week voyage across the Atlantic that included only two deaths, a period called the "great sickness" and "starving times" began- the mortality rate for the first eight months of settlement was at almost fifty percent. This figure includes one human spontaneous abortion that occurred about six weeks after landing was made on Cape Cod.[281] Eight years later, history seemed to repeat itself when the Puritan settlers landed nearby in what is now Salem, Massachusetts.

The landscape they encountered would have been very similar to one that the Pilgrims had encountered earlier-- a well oaked forest, abounding with deer, interspersed with some stretches of cleared land that had been

## DISGUISED AS THE DEVIL

depopulated by the waves of sickness that had decimated the local Native American tribes. Once again, cleared land at the edge of the forest had accumulated several years' worth of shrubby growth. Once again, there would have been a decrease in the seasonal burning of undergrowth. Once again, the settlers became very, very sick. Although a handful of stray settlers predated them, the Massachusetts Bay colony was to a large degree initially settled by two fleets. It is estimated that eighty colonists out of the two hundred from the first 1628-29 Higginson fleet died and many became sick. More than two hundred out of nine hundred colonists from the second 1630 Winthrop fleet died, and one hundred went back to England where most of them died anyway, making for a +/- thirty-three percent initial death rate.[282]

    At Plymouth, the mortality was blamed on cold weather, catching cold, poor living conditions, poor diet and scurvy.[283] In Salem, Boston, and Charlestown it was blamed on hot weather, fever, the mental defect of "pining for home in England," poor living conditions, poor diet and scurvy.[284] When population groups went to Connecticut, they also seem to have once again encountered an initial phase of sickness. It also infected their cattle, causing malaise and spontaneous abortion.[285]

    The diagnosis put forward by William Bradford, Edward Winslow, Plymouth physician Samuel Fuller, and John Winthrop in their writings is that the initial sickness in Plymouth and Massachusetts Bay was scurvy caused by a diet that lacked grains and other proper food. Some of the symptoms pointing to this were: lameness, joint pain, and extreme fatigue to the point of not being able to get out of bed. This story was repeated by William Wood who wrote, "whereas many died at the beginning of the plantations it was not because the county was unhealthful but because their bodies were corrupted by sea diet." Because of the great sickness, John Winthrop sent a ship back to England to bring over an emergency supply of proper English food and lemon juice (a known cure for scurvy). The level of scurvy does seem to have abated by the time the ship returned in February of 1631. Credit was given to the lemon juice, not the time of year (ticks are not active in the coldest part of the winter) for causing a cessation of new cases of 'scurvy.'[286] At face value, on arrival after a long period at sea, some passengers may have had a deficiency of vitamin C and developed scurvy. This self-diagnosis becomes suspect, however, because once they arrived on land the settlers immediately began to supplement their shipboard diet with indigenous foods and mention doing so. Clams, mussels, oysters, wild onions and leeks, and especially rose (hips) are mentioned. Scurvy grass grew in salt marsh areas.[287] A diet that included even some of these food items would actually have been extremely rich in vitamin C.

The modern recommended daily adult allowance of vitamin C is 60 mg.[288] This is a fairly general number meant to cover the nutritional needs of a wide spectrum of body weights. The number or seventeenth-century adults can only be estimated but even a small amount of some of the foods they mention eating would have provided an adequate daily allowance for vitamin C.

Clams, which were described as "being at their doorstep," [289]would be an excellent source of vitamin C. A modern size clam contains 8.84 mg of vitamin C.[290] To mentally picture this; three modern soft-shell clams *(mya arenaria)* contain vitamin C equal to the amount found in one lime. However, shellfish were described as being much larger then than they are today. Clams were "as big as a penny white loaf" of bread. Oysters were "great ones in the form of a shoe horn," some "a foot long." Frances Higginson described oysters at Salem that were ten inches wide.[291] This larger size at contact is supported by archaeological evidence.[292] If a small modern oyster contains 1.78 mg,[293] a ten-inch oyster might outstrip even a lime for vitamin C content.

Onions may still be a common ingredient in many recipes today because, in addition to adding taste, they add vitamin C and are highly antimicrobial. Wild onions and leeks were described as being available even in the winter (in 2003, for example, they were still found in edible condition growing on Cape Cod on November 12).[294] When this ingredient was added to pottage for taste, it made it into an excellent source of vitamin C. Wild onions contain 38 mg in every ounce. Wild leeks weigh in at 80 mg for every 3.5 ounces.[295]

While they certainly believed that they had scurvy themselves, the scurvy diagnosis for the Winthrop fleet in Salem and Boston is even more suspect, because of the time of year and one of the first activities they mention doing after landing. When one ship arrived on June 12, the passengers immediately picked (and presumably ate) strawberries. These strawberries were described as "in abundance, very large ones, some being two inches about," so prolific that "one may gather half a bushel in a forenoon."[296] A strawberry is a vitamin C powerhouse with a whopping 86.18 mg per cup.[297] Again the scurvy diagnosis does not match with the facts. It is also difficult to believe that in a countryside full of rose hips (a known anti-scorbutic in the seventeenth century) and sassafras that was so plentiful that they were exporting it back to England, they would not have used them to help cure their own illnesses. The general population

## DISGUISED AS THE DEVIL

was certainly receptive to ingesting medicinal fluids to help their disorder. Nicholas Knopp, for example, was found guilty after he was brought before the Massachusetts General Court in 1631 for taking it upon himself to cure the scurvy with a water of no worth nor value, which he sold at a very dear rate. It is very likely that the sick settlers tried traditional vitamin C rich "cures" but that they did not work.[298]

**ENGLISH SETTLEMENT**

252 Author unknown, historically attributed to Mother Goose.
253 A.R. Hall, A.K.G. Jones and H.K. Kenward, "Cereal bran and human faecal remains from archaeological

# ENGLISH SETTLEMENT

deposits- some preliminary observations," *Site Environment and Economy- British Archeological Reports* (1983), 85-104.
254 James Baker, "17th Century Yeoman Foodways," in *The Plimouth Plantation Cookbook* (Massachusetts: Plimouth Plantation, 1990), 5-13.
255 Clarence Tabor and Associates, *Tabor's Cyclopedic Medical Dictionary* (Philadelphia: F.A. Davis Company, 1952), S-27. This was the oldest pre-discovery of the Lyme disease spirochete edition that I could get my hands on. It defines scurvy as a disease characterized by mucosa hemorrhage, prostration, and ecchymoses -a form of macula (spot, blemish or stain) appearing in large irregular shaped hemorrhagic areas of the skin. This description is similar to modern descriptions of sore throat, fatigue and various rashes and skin reactions in Lyme disease. People with certain genetic make ups may be more susceptible to scurvy: see De Buyzere, Marc L., Langlois, Michel R., Torck, Mathieu A., and Delanghe, Joris R, "Vitamin C Deficiency and Scurvy Are Not Only a Dietary Problem but are Codetermined by the Haptoglobin Polymorphism," *Clinical Chemistry* (8/1/2007).
256 Zita Weise Prinzo, *Scurvy and its prevention and control in major emergencies,* (World Health Organization, 1999).
257 Watt, J., E. J. Freeman, and W. F. Bynum. *Starving Sailors: The Influence of Nutrition upon Naval and Maritime History* (London: Nat. Maritime Museum, 1981).
258 Prinzo, *Scurvy.*
259 See Kumarave Rajakumar, "Infantile Scurvy: A Historical Perspective," *Pediatrics* 108 No.4 (October 2001) for a good overview of the history of the development of treatments for scurvy.
260 Edward Winslow, *Mourt's Relation* (Bedford, Massachusetts: Applewood Books, 1963), 24.
261 William Bradford, *Of Plymouth Plantation 1620-1647* (New York: The Modern Library, 1981), 85 also see Chapters 9-13.
262 Stephen Mennell, *All Manner of Food-Eating and Taste in England and France from the Middle Ages to the Present* (Urbana and Chicago: University of Illinois Press, 1996), 40-43.
263 Sandra Oliver, *Saltwater Foodways* (Mystic, Connecticut: Mystic Seaport Museum, Inc., 1995), 79.
264 John Smith, *General Historie of Virginia, New England and the Summer Isles,* (London, 1624). Reprinted in Philip Barbour, ed., "Chapter 2, What Happened till the First Supply," Complete Works Captain John Smith. (Chapel Hill: UNC Press,1986), 2.
265 Bradford, Plymouth, 10.
266 Michael P. Ricards, Rick Schulting and Robert E. M. Hedges, "Sharp Shift in diet at onset of Neolithic," *Nature* 425 (September 25, 2003), 366.
267 Bradford, *Plymouth*, 122.
268 Edward Winslow, *Mourt's*, 67.
269 Oliver, *Saltwater*, 332.
270 Gregory J. Brown and Joanne Bowen, "Animal Bones from the Cross Street Back Lot Privy," in Ronald L. Michael, ed., *Historical Archeology* 32 No.3 (1998), 75.
271 Smith, *Historie*, from a list of the names of "them that were the first planters," 47 out of 100 are listed as occupation: Gentlemen.
272 Charles D. Cheek, "Massachusetts Bay foodways: Regional and Class influence," Ronald L. Michael, ed., *Historical Archeology* 32 No.3 (1998), 153-161.
273 John Winthrop, *The Journal of John Winthrop 1630-1649* (Cambridge, Massachusetts: Belknap Press, 1996), 23.
274 John Winthrop quoted in David Hackett Fisher, *Albion's Seed: Four British Folkways in America* (New York: Oxford University Press, 1991), 137.
275 Cheek, *Foodways*, 153-161.
276 Winslow, *Mourt's*, 23.
277 Bradford, *Plymouth*, 135-136.
278 Bradford, Ibid, 144.
279 Robert Atkins, *Dr. Atkins New Diet Revolution* (N. Y: Avon Books, 2002),47-55.
280 Bradford, Plymouth, 144
281 Eugene Stratton, Plymouth Colony, Winslow, *Mourt*'s, 23.
282 Bradford, Plymouth, 85-86, Winthrop *Journal,* 45, 200.
283 Winthrop, Ibid.
284 Winthrop, Ibid.
285 William Wood, *New England's Prospect* (University of Massachusetts Press, Amherst, 1977), 57.
286 John Winthrop, Jr. went back to England for this emergency supply run. He returned in February of 1631.

287 Bradford, *Plymouth*, 74,100, Winslow, Mourt's, 21, 39, 84, Winthrop, *Journal*, 35, Wood *New England*, 36-37, 50, 55-57.
288 United States Department of Agriculture, *Vitamin C Content in Foods Nutrient Database*, Release 12 (1998).
289 United States Department of Agriculture, Ibid.
290 Wood, *New England Prospect.*
*291 Frances Higginson in Everett Emerson, Ed., Letters* from Massachusetts Bay Colony 1629-1638 (University of Massachusetts, Amherst, 1976), 12-24.
292 For examples, see shells in the Peabody Collection at USM Archeological lab. USDA, Vitamin C.
293 USDA, *Vitamin C.*
294 Personal conversation with National Park Service in Eastham, Massachusetts, 11/12/2003, see also Paul Sherman and Jennifer Billing, "Darwinian Gastronomy: Why we use spices," *BioScience* Vol.49 No.6, (1999), 456.
295 Thomas Zennie and C. Dwayne Ogzewalla, "Ascorbic Acid and Vitamin A content of Edible Wild Plants," *Journal of Economic Botany* (1977), 76-79.USDA, Vitamin C.
296 See Thinking in *Time, Wood, Prospect, 36.*
*297 USDA*, Vitamin C.
301 Nathaniel Shurtleff, *Records of the Governor and Company of the Massachusetts Bay in New England: Volume 1* (New York: AMS Press, 1968), 83.

## IV. WAS IT LYME DISEASE?

"The pathogens were always here, but we had no vectors that fit both the reservoir and the people. That was accomplished when the deer herds came back."[299]
Andrew Spielman M.D.

Nicholas Knopp may have been an innocent man. In the absence of lemon grass and shepherds purse, he may have been offering an elixir full of vitamin C that would have cured even the scurviest of those with C deficiencies. However, it did not work because the settlers did not suffer from a lack of vitamin C. The colonists suffered from an ailment that had nothing to do with their diet. What they really may have needed were elixirs that contained antibiotics that would cure the bacterial infection of *Borrelia burgdorferi*. Just as many descendants of those first English settlers still inhabit the New England landscape, there are other descendants of other organisms from the seventeenth century that still live here today, including white tailed deer, ticks and even bacteria. Massachusetts had thriving deer herds in the seventeenth century.

Arriving on November 11, 1620, at what would have been a time period for questing adult *Ixodes scapularis* ticks, the Pilgrims violated each and every anti-tick recommendation put forth by modern epidemiologists. The historic records state that initially the weather was quite warm and pleasant which means that any local ticks could have been active. The Pilgrim's methods of exploration brought them into direct contact with a mosaiced landscape full of marginal areas-'the geography of risk' for Lyme disease. The men describe exploratory journeys in which they marched through boughs, bushes, and edge ecological zones full of "brush, wood-gaille, and long grass." They followed paths made by deer.[300] Deer have tarsal glands halfway up their legs that produce pheromones to mark their territory as they walk. Ticks find this gland secretion highly attractive, which leads to a greater abundance of ticks questing along deer traveled paths. By using deer paths, these explorers not only exposed themselves to a greater concentration of questing ticks, they probably acquired an odor that would attract ticks.[301]

## DISGUISED AS THE DEVIL

When the first group of men left the *Mayflower* to explore the landscape of Cape Cod, they would have been wearing many layers of clothing. For the upper body, a man in the seventeenth century wore a long, short-sleeved collared linen shirt. On top of that, he wore a doublet, which was relatively close fitting with long sleeves and padded shoulders. It buttoned down the front. A collar and cuffs might have also been attached. A felt hat or cap was usually worn and in cold weather, a cloak was added for warmth. For the lower body, men wore front buttoning breeches, which extended to the knee and were worn with knee length socks and low-heeled leather shoes or boots.[302]

Preparing to face the unknown dangers of 1620, the exploring men were all armed and wore protective chain mail under a form of upper body armor called brigandine corselets. These were made of quilted cloth with small metal plates sewn inside. William Bradford noted that the boughs and bushes in some areas were so thick that they tore their "armour in pieces."[303] Miles Standish, as a professional soldier and veteran of the wars in the Netherlands, probably owned his own set of full metal gear with a metal helmet. He was also equipped with the most modern of seventeenth-century weapons: the revolutionary snapchance. This was the first firearm that used a flint to spark a firing mechanism. Others carried cumbersome, difficult to aim, matchlock muskets that required a lit wick in order to be fired.[304]

While this male apparel, with its layering and pants tucked into tall stockings, may have been mildly protective (although once covered with deer pheromones this advantage may have been lost) if a tick found its way to human flesh it was probably there to stay unless scratched off. Bathing was not part of any prevailing practices at the time.[305] A lack of bathing would tend to be protective of any attached ticks. A tick needs to bite, attach, and be in contact with a host's blood for at least 24 (some experts say 36 some say more) hours to successfully transfer the *Borrelia* bacteria and subsequent Lyme infection. If the attachment lasts at least that long the infection rate can be extremely high.[306]

The apparel of the women who had left the *Mayflower* to refresh themselves would have put them at high risk for Lyme disease. A woman's one undergarment consisted of a long short-sleeved linen shirt that was fastened in front. Over this shirt, one or more ankle length petticoats were worn. Over these, she wore a wool gown that consisted of two parts: a bodice (which may have had removable long sleeves) and a skirt. The skirt was ankle length. Women wore knee high socks held up by garters, low shoes, and a cloak in cold weather. If the children were

## WAS IT LYME DISEASE ?

allowed on land, many of them would also have been wearing long skirts. Children of both sexes wore long gowns made of linen or wool with long sleeved bodices until they were about 8 years old.[307]

In November of 1620, some women came ashore on Cape Cod to wash, and in the process did what could be best described as an unintentional tick collecting procedure. In modern epidemiology, this is called a flag or drag test. It is done by dragging a light colored cloth over the ground and then counting the number of ticks that have attached themselves to the cloth.[308] The Pilgrim women's clothing left them vulnerable to collecting questing ticks when their long cloth skirts were dragged over the ground. Any ticks attached to their skirts would have been carried unnoticed aboard ship. Once on a skirt hem, a tick could easily find exposed flesh for a blood meal.

Even worse, piles of wood and shrubby possibly tick filled juniper branches were collected and transported from shore to ship to be burned on a daily basis.[309] A 1999 study of the relative potential for acquiring *Ixodes scapularis* tick nymphs while crawling, walking, or sitting in a deciduous woodland found some risk for acquiring ticks existed for all three activities, but that crawling on the ground raised the risk. In addition, tick nymphs were found on +/-85 percent of the logs large enough to sit on in the study area, which indicates that sitting on one was tick-risky behavior. While it is impossible to know, it is highly likely that both men and women sat on convenient logs in 1620.[310] The Pilgrim men walked along deer traveled paths and then slept overnight in the grass at the edge of the woods. Because of these activities, if there were ticks infected with Lyme disease in the area in 1620, they would have had numerous opportunities to infect both the settlers and the *Mayflower* itself.

Fig. 26. *Drag test.* USDA

Only after contact with the landscape of Massachusetts, people began to get sick. Some of the symptoms of that William Bradford and Edward Winslow described were flu like symptoms including coughing (scarce any of us were free from vehement coughs),[311] headaches (he complained greatly of his head),[312] joint pain (William Bradford feels pains in his ankles and hip),[313] fatigue (the weakness of our people),[314] death, and possibly a congenital transfer of the bacteria to a fetus. Goodwife Allerton delivered a stillborn child on December 22 and then also died herself within a few weeks.[315] Out of 102 settlers and an unknown number of ship's crewmates, only 52 settlers and a handful of

## DISGUISED AS THE DEVIL

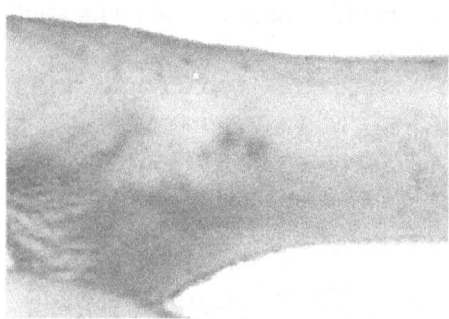

**Figure 27.** *Photograph of scurvy macula on an ankle: note red spot and compare with ECM.*[316]

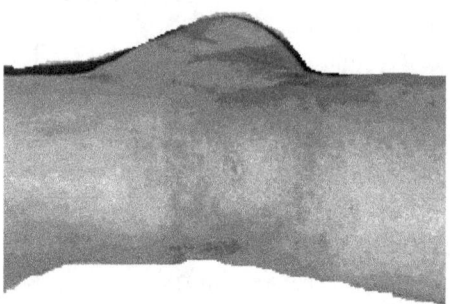

**Figure 28.** *Photograph of a Lyme ECM rash on an ankle: note red spot.* www.hvcn.org

crewmates survived through that first winter. The oldest man, James Chilton, was one of the first to die. Many other men, most of the women, and many of the children died. Others got very sick. William Bradford himself was "vehemently taken with a grief and pain, and so shot to the hackle-bone (hip)." He felt some pain in his ankles "by times" and came close to dying but did survive.[317]

All of these symptoms would be highly similar to those of scurvy, which would account for the original diagnosis. The Pilgrims were not in the middle of a witch inquiry in 1620 so there are no detailed bodily examinations for marks included in the historical records from that time. Although no skin rashes were noted by Bradford, it is interesting to note that the 'bull's eye' rash of Lyme disease could easily have been mistaken for the skin spots that accompany scurvy. The fact that the contact disease was diagnosed as scurvy may implicitly suggest the

## WAS IT LYME DISEASE ?

appearance of some form of skin rash. The scurvy symptoms of bleeding gums, foul breath and loosened teeth are also not mentioned anywhere.

There is also an odd distribution of survivors that might argue against contagious diseases that most closely mimic Lyme disease symptoms-especially *streptococcus* infective rheumatic fever. Prior to the discovery of the *Borrelia burgdorferi* spirochete, many people who suffered from Lyme disease were sometimes given some sort of rheumatic diagnosis. One of the factors that led to the discovery of the Lyme disease spirochete was the fact that clusters of arthritis began to occur in specific geographic areas-like Lyme, Connecticut. Since rheumatoid arthritis is considered to be individualized immune response, not a contagious disease, it made the scientists at Yale University begin to look for another causative factor for the clusters of infection. Eventually *Borrelia burgdorferi* was discovered.[318]

The precursor of rheumatic or scarlet fever is usually strep throat, a common infection that is caused by the *streptococcus* bacterium. This bacterium has been one of the most extensively studied and analyzed pathogens known to mankind. After a short incubation period,[319] it can spread rapidly in crowded conditions. It is spread by person-to-person contact with infected saliva and mucus. In some cases, a streptococcal throat infection leads to a secondary disease- rheumatic fever-which may or may not have associated symptoms of rheumatoid arthritis. The arthritic form of the disease is thought to be an individualized immune response to the infection itself. The mechanism for rheumatic disease development within an individual patient's body is still unknown. However, it appears that if the Pilgrims had been heavily infected with *streptococcus,* it would have stuck earlier during their crowded cross-ocean journey, which was instead disease free. [320]

In the aftermath of rheumatic fever, victims suffer profound life altering heart and kidney problems.[321][The carditis and other heart complications associated with Lyme disease, however, are thought to resolve themselves over time] There is no record of the population at Plymouth suffering from the kind of debilitating problems that are the lingering hallmark of rheumatic fever. To the contrary, those who survived the contact "sickness" lived long lives, outstripping the life spans of their contemporaries back in England by decades.[322]

Some members of the *Mayflower* population who would normally be at the highest risk for any type of contagious infection survived. They included some of the weakest, the oldest, and the youngest-newborns. These survivors have a common attribute: they were among the least

## DISGUISED AS THE DEVIL

likely to go ashore. Survivors included the two women who had just given birth prior to contact with the land, as well as their newborn children. William Brewster and his wife, almost the most elderly in the group, survived- he did not even get sick although he spent his time nursing others. Miles Standish, who had a background of participating in military campaigns in rugged areas of mainland Europe and who probably tromped his way through the countryside in front of the other exploring men clad in protective full military apparel (it would be difficult for a tick to attach to a metal surface) was likewise untouched by illness. He also spent his time nursing the ill back to health without catching the disease. This argues against an infection spread by casual person-to-person contact.[323]

    The various members of this population may have spent most of their long journey across the ocean below deck and out of the sun. And even when they went above deck, the sun's angle at the northern latitude where they were traveling did not provide much in the way of ultraviolet B rays-the kind needed to produce vitamin D in the skin-at that time of the year. A low level of vitamin D at contact may have made these settlers more susceptible to a variety of diseases, including Lyme disease. Prior to emigration, most of the pilgrims had been involved in various occupations involving the manufacture of cloth-jobs that were often done indoors. Miles Standish may have been an exception because he was a professional soldier. A pale complexion was considered to be "fresh" and gentile while a tanned skin was the mark of a farmer, laborer or soldier in the seventeenth century. William Bradford noted that after reaching the New World the pilgrims eventually acquired suntans. He wrote, "their labors abroad had abated the freshness of their former complexion."[324]

    Hormonal secretions related to stress may also have a triggering affect on the *Borrelia burgdorferi* bacteria.[325] When exposure to Lyme disease was added to the stressors of extremely crowded living conditions, exposure to the elements, poor shelter, vitamin D deficiency and fear, it produced a fatal encounter. Contact with the marginal, brushy, edge terrain, the 'geography of risk' for Lyme disease, at a peak time period for questing ticks could and did make people very sick. This was the result of a series of events that had occurred prior to 1620. The environment of Cape Cod, Plymouth and Boston area had been undergoing changes in response to the dramatic decline of the Native American.

Edward Winslow described the countryside: "Thousands of men have lived here, which died in a great plague not long since... as we passed along, we observed that there were few places by the river but had been inhabited, by reason where-of much ground was clear save of weeds which grew higher than our heads. There is much good timber, both oak, walnut tree, fir, beech and exceedingly great chestnut trees...though the country be wild and overgrown with woods, yet the trees stand not thick, but a man may well ride a horse amongst them."[326]

Visiting a Native American village, Winslow was offered some acorns to eat. Acorns were one of the staple foods of these indigenous people. They provided a reliable and nutritious source of food. A native family, it has been estimated, could harvest and eat over a half a ton of acorns in a year's time. When the Native human population in New England was decimated by disease, it created a bonanza of acorns for other species. Deer, turkeys,[327] bear and passenger pigeons all consumed acorns. The deer population would also have increased because there was a decline in Native hunters.

New England's seventeenth-century tick populations would have thrived in response to this increase in the white tailed deer population. Although they do not transmit the *Bb* bacteria, deer are an important maintenance host for *Ixodes scapularis* tick populations, especially in the adult stage. A modern study done at Great Island in West Yarmouth, Massachusetts points out this close relationship between deer and tick populations. A tick survey, using the 'drag test' method, was done between 1982 and 1985 on this small island off the coast of Cape Cod that was populated by white tailed deer. The results showed both a high level of deer ticks and a high Lyme disease infection rate among the collected ticks. The Island was privately owned and the owner, whose grandchild had contracted Lyme disease, was in no mood for unequivocal results. After what must have been a nasty slaughter it was estimated that ninety percent of the island's deer population had been eliminated. Several subsequent tick tests on the Island show a corresponding drastic reduction in the *Ixodes* tick population.[328] The same thing happened when deer were eliminated from Monhegan Island, off the coast of Maine.[329]

## DISGUISED AS THE DEVIL

## WAS IT LYME DISEASE?

299 Peter Wehrwein, "Nantucket Fever: Entomologist Andy Spielman's Search for the Creeping Carrier of Babesiosis, Ehrlichiosis and Lyme disease," *The Harvard Public Health Review* at www.hsph.harvard.edu.
300 William Bradford, *Plymouth*, 73-78 and Edward Winslow, *Mourt's Relation*, 18-29.
301 John Carroll, "Kairomonal activity of white tailed deer metatarsal gland substance: a more sensitive bioassay using Ixodes scapularis," *Journal of Medical Entomology* 35 (1996), 90-93.
302 Herbert Norris, *Tudor Costume and Fashion* (New York: Dover Publishers, 1997), 46-66.
303 James Deetz and Patricia Scott Deetz, *The Times of Their Lives* (New York: Anchor Books, 2000), 246-247.
304 Bradford, *Plymouth*, 77 and Winslow, *Mourt's*, 35-36.
305 See Internet Medieval Sourcebook at www.Fordham.edu for bathing practices.
306 Karen Vanderhoof-Forschner, *Everything You Need To Know about Lyme Disease and Other Tick Borne Disorders* (New York: John Wiley & Sons, 1997).
307 See Janet Arnold, *Patterns of Fashion 1660-1860* (U.K.: Drama Publishers, 1985) for information about the cut and construction of these "dresses."
308 Karen Vanderhoof-Forschner, Flag test, *Everything*, 136-37.
309 Winslow, *Mourt's*, 19.
310 The Pilgrims probably spent time doing various activities that brought them into contacts with ticks during their stay on Cape Cod. See John Carroll, "Relative Potential for Acquiring Nymphs of Ixodes Scapularis while walking, crawling or sitting in a Deciduous Woodland," TEKTRAN United States Department of Agriculture- *Agriculture Research Service News* (September 10, 1999) archives at www.nal.usda.gov.
311 Winslow, *Mourt's*, 30.
312 Bradford, *Plymouth*, 95.
313 Winslow, *Mourt's*, 45.
314 Bradford, *Plymouth*, 105.
296.Altaie, S., S. Mookherjee, E, Assian, F.Al-Taie, S.M. Nakeeb, S. Siddiqui and L.Duffy. "Transmission of Borrelia Burgdorferi from experimentally infected mating pairs to offspring in a murine model," VII *International Congress on Lyme Borreliosis Report*,1996.
316 This picture is from the well worn medical book, the 1926 *Mothers Home Companion Medical Guide* that I grew up with; scurvy does not appear in modern home medical guides.
317 Winslow, *Mourt's*, 24.
318 Alan Barbour and Durland Fish, "The Biological and Social Phenomenon of Lyme disease," *Science* 260 (June 11, 1993), 1611.
319 1 to 4 days in most cases.
320 Rheumatic Fever Information sheet-www.sd.state.gov/doh/Pubs/rheumat.htm. See also O'Loughlin RE, Roberson A, Cieslak PR, Lynfield R, Gershman K, Craig A, et al. "The epidemiology of invasive group A streptococcal infection and potential vaccine implications: United States, 2000–2004." *Clin Infect Dis.* 2007;45:853–62.
321 Leon Gordis, *Epidemiology* (W.B. Saunders Company, New York 2000),11-12, also 41.
322 Bradford, *Plymouth*, 364. They did, however, sometimes suffer from arthritis and dementia. Samuel Sewell once went on a trip to meet one of these arthritic "ancients" who was still alive 70 years later when Plymouth Colony was joined to Massachusetts.
323 George F. Willison, *Saints and Strangers* (Orleans, Massachusetts: Parnassus Imprints, Inc, 1945),166-68.
324 Bradford, *Plymouth*, 144.
325 Denise Lang, *Coping With Lyme Disease* (New York: Henry Holt, 2004), 138.
326 Winslow, *Mourt's*, 51.
327 In Lane, R.S., et al., "Wild Turkey(Meleagris gallopavo) as a host of Ixodid ticks, lice and Lyme disease spirochetes" in *California State Parks* May, 2007, a study done by the University of California at Berkeley found that 44.2 % of turkeys were infested with ticks and can be considered to be important vector hosts for the tick but not for the Bb bacteria.
328 Howard Ginsburg, ed., *Ecology and Environmental Management of Lyme disease* (New Brunswick, N.J.: Rutgers University Press, 1993), 148-149.
329 Personal conversation with Charles Lubelczyk of the Vector Borne Disease Research Laboratory at Maine Medical Center in Portland, October 17, 2004.

SCURVY DÉJÀ VU?

# V. THE MASSACHUSETTS BAY COLONY: SCURVY DÉJÀ VU?

Eight years after the Pilgrims experienced what they thought was 'scurvy,' the next group of settlers to come to Massachusetts experienced similar symptoms. Once again, their self-diagnosis was scurvy. Beginning with their first settlers in Salem in 1628, the Puritan settlements around Boston, less than forty miles north of Plymouth, experienced their own initial period of sickness. On March 15,1630, John Pond wrote: "People here are subject to disease for here have died of the scurvy and of the burning fever two hundred and odd, besides many layeth lame, and all Sudbury men are dead but three and the women and some children."[330] Sudbury was even closer to the deer and tick infested forest edges than Boston.

    The Winthrop Fleet arrived in the spring of 1630 and some of its passengers spent time exploring the Massachusetts Bay region. This is a high-level time period for questing tick nymphs. John Winthrop and others went on exploratory journeys while trying to decide where to settle. The Puritans changed their minds about their chosen settlement site in response to a rumored threat from the French. This required some degree of trekking back and forth through an overgrown landscape. They were then subjected to a period of disease. Once again, interaction with the landscape immediately proceeded the period of sickness. Even the Puritans themselves made comparisons with the experience of Plymouth as a model to help get them through a similarly difficult time. John Winthrop wrote, "the first at Plymouth suffered a similar tempest of pestilence but have now grown healthy and thriving." [331]

    Traditional and folkloric medical cures may hold some clues to the root causes of some diseases in the seventeenth century. When we conceptualize scurvy in modern medical terms it is as a disease of vitamin C deficiency. The scurvy diagnosis in Boston, Salem, and Plymouth may have been an accurate diagnosis by that century's standards, because the

seventeenth century it was a more complex picture with treatments for two causative factors lumped together as one.

The form of illness that the English called land scurvy in the seventeenth century may have sometimes been the bacterial infection of Lyme disease or/in combination with/ a springtime vitamin C deficiency. There may be some interrelationship that adds to an already complex situation because at least one modern physician has stated that she sees "that [scurvy] in a lot of patients with spirochetal illnesses." [332] The modern definition of scurvy as a disease caused by a deficiency of vitamin C is the result of scientific work done after The Enlightenment. This deficiency only became prevalent in the seventeenth and early eighteenth century, when sailing boats began to consistently make journeys that lasted more than three months-the length of time it takes for scurvy to develop in a vitamin C deprived dietary environment. Most traditional cures for scurvy, like lemons, limes and rose hips, contain high levels of vitamin C. Other English folkloric herbal remedies for scurvy, however, like lemongrass and Shepherd's purse have antibacterial properties. In a study of its antimicrobial properties, lemon grass was found to inhibit or kill about 85 percent of the bacteria it was put into contact with- and while Shepard's purse also contains vitamin C, the conflict and combination of differing cures with good folkloric reputations may lie in two roots for what was considered the same disease.[333] Any folkloric cure was probably preferable to the common and extremely painful medical cure practiced in that era: bleeding.

When John Winthrop sent in 1630 for Plymouth's Doctor Samuel Fuller "who did such good work with the scurvy"[334] to help his sick colony it may have been a mistake. Dr. Fuller's cure, in modern times known to aggravate many conditions, consisted of "bleeding" the afflicted. His actions may have actually elevated the death rates in both colonies. [335] Another wave of sickness corresponds to the move by a large segment of the population from the outlying settlements in Massachusetts Bay Colony to Connecticut during the mid 1630's. This was partially the result of problems that they had with the health of their cows. Once again, they looked at the quality of food as a causative factor- in this case the grass eaten by their cattle. In their ethno-centric minds, even English grass was better than American grass for their cattle. Fields "that were grubbed down to bare soil by pigs and goats" could then be made superior by being sown with English grass seed.[336] Inhabitants of outlying edge areas like Newtown, Dorchester, Salem Village and Concord felt that

## SCURVY DÉJÀ VU?

wild American grass was making their cows sick by being "too rich" and chose to move to the greener pastures in Connecticut.[337]

The symptoms described (weakness, failure to produce milk, spontaneous abortion) have all the strong earmarks of bovine tick borne infestations. In Concord, the problem with cows was compounded by a frustrating problem with the high population of wolves. In addition to killing a stray calf or two, every time the settlers planted a field of corn with fish as fertilizer, the wolves would come at night, dig everything up, and eat the fish.[338]

Things, however, did not initially go too well once the settlers moved to Connecticut either. The settlement of Connecticut after the Pequot War is an interesting case: Lyme was founded during this wave of population movement. The first settlers moved from Massachusetts to Connecticut by water- sailing around Cape Cod and then up the Connecticut River to points near present day Hartford. John Winthrop notes, however, that soon after the first settlement, an overland route was found between the two areas and a second wave of outward expansion was made using this path, including a disastrous cattle drive. Many of the cattle sickened, refused to cross the last river, and were left on the distant side where many succumbed to harsh winter conditions. Those that were left alive were sick. In 1636 John Winthrop noted, "Their cattle died, many of them, and cast their young, as they had the year before."[339]

This experience is strikingly similar to the descriptions of problems experienced during early cattle drives between Texas and California. The greatest scourge of disease among cattle herds in the nineteenth century was called Texas or Southern fever. The disease was attributed to various factors and the attempts of early cattlemen to protect their herds were futile. By the 1880's, observation and deduction had shown that the fever was usually contracted in the aftermath of contact with other herds. In addition, it was found that not only contact, but the mere passage over a pathway recently trod by perfectly healthy cattle, or pasturage where they had lately been often resulted in contraction of the disease. Cattlemen reported the disease was prevalent from July to December or whenever the date of the first frost occurred. Richard Gird expressed the bewilderment of the cattlemen when he wrote in his journal in 1881:

> Cattle still continue dying. It is certainly very hard to get at the cause or a cure. Post-mortem of many show many organs affected.

On autopsy, some of the cattle's organs were found to have turned black.[340] In 1888 the cause of the disease was discovered to be "an

## DISGUISED AS THE DEVIL

intracorpuscular parasite" that is now known as babesiosis. During the next year investigations proved that the transmitting agent was a tick. Research soon proved that ticks alone, scattered on pathways and pastures that were later used by healthy cattle invariably resulted in an outbreak of the disease. Similar to Lyme disease, the bite of a tick was proved to be essential for the transmission of Texas fever. Babesiosis is a co-infectant carried by the *Ixodes* tick in modern New England.[341]

A description of the forested, shrubby landscape that was traversed during that first seventeenth century overland cattle drive between Massachusetts and Connecticut shows that they were traversing the pathways and edge areas that were "abound" with white tailed deer in the seventeenth century, that the drive was made at the height of adult tick questing season, and, like the cattle in California, the settler's cattle became diseased soon afterwards. These similar illnesses may have had a similar causative agent.

After the initial phase of exploration and movement and until population growth began to press against the 'wilderness,' contact with marginal areas in the environment became a sporadic occurrence with only sporadic sickness occurring. New arrivals to Plymouth and Massachusetts Bay did not suffer from any subsequent initial periods of great sickness or 'scurvy.' Even after suffering from the effects of contact disease, both the men and those women who survived complications from childbirth (the leading cause of death for women in the seventeenth century) were usually able to live long lives. Life spans of sixty, seventy, and even eighty years were reported. At the same time, the expected lifespan in England actually decreased from 36 years in 1620 to 35 years in 1680.[342]

Once a site for settlement was decided on, the early settlers proceeded to burn off undergrowth and leaf litter, set loose their pigs on the forest edges to gobble up acorns, and place a physical and mental overlay of English social and civil order onto the landscape. Salem, Plymouth and Boston were all set up initially with a central nucleus of house lots with a set of common fields laid out around it. Traditional husbandry practices were altered to fit the New England climate. Animals needed to be protected from wolves and were more likely to be sheltered during the winter than they had been in England. New American crops like pumpkins, beans and corn were grown. Cornfields were fertilized with fish, an adoption of a Native American practice. English grass seed was imported. Like the human settlers, English grass thrived in New

## SCURVY DÉJÀ VU?

England and it begins to show up in bog core samples almost immediately after 1620.[343]

New England's almost instantaneous creation of a thriving local market for cows and swine was the foundation for Massachusetts's fledgling economy and its first West Indian links. The domestic animal and meat market kept New England afloat during the 1630's and offered a sure commodity for the West Indian trade when it began to develop in the 1640's. By 1633, John Winthrop stated that the domestic cattle trade had saved the Massachusetts economy from ruin. The colony "by reason of so many foreign commodities expended, could not have subsisted to this time [except] that it was supplied by cattle and corn, which were sold at a very dear rate." Cows sold for between twenty and twenty six pounds. These high prices led to the importation of animals from both England and Virginia for profit.[344]

The introduction of an English style economy had some immediate effect on the forest landscape, especially on the home ranges of deer, mouse and wolf habitat occupants. A landscape model for seventeenth-century Massachusetts would show a high pre-contact deer population that grows even higher as their prime predators diminish over time. It would show a wolf population that is affected and, by the nineteenth century, eliminated by a bounty eradication system. [345]

There are several important factors that influence the development of this model. The first factor is the seventeenth century elimination of one of the deer's primary predators- the Native American. During the early seventeenth century, it is estimated that epidemics of smallpox and the plague killed nearly ninety percent of the Native population in what is now the state of Massachusetts. This should be reflected on a population chart by a decrease in the number of hunters, and a reactive increase in the total deer population. With more deer available, there would have been another reflective increase- in the number of predators.

"Lions" were seen and described from 1620 on. In Plymouth, the first settlers heard the growls of a mountain lion and had to hold their dogs back from attacking it.[346] Wolves were considered flat out pests. They became the pariah of the wilderness--dark, insidious predators biting at the heels of civilization. They had a price on their heads from almost the moment of contact with the English colonists. Well nourished on deer meat, this thriving wolf population was unfortunately not discerning enough to know a domesticated animal from their wild prey. When they began to add pork, beef, and mutton to their diet, it was not tolerated.[347] In 1678 Salem Village was rimmed by a set of wolf traps.[348]

## DISGUISED AS THE DEVIL

The last wolf bounty in Massachusetts was paid in the nineteenth century at the end of a successful eradication program that took over 200 years to complete.[349]

A second factor, perhaps unique to the English colonists in Massachusetts, was the almost complete lack of interest in hunting deer that was displayed by the almost homogeneous middle class group of settlers who went there. In English common law, game (especially deer) was by custom owned by the crown or upper class nobility who had the legal right to hunt their own animals. Most of the first settlers in Massachusetts did not come from that social class. These settlers had not been deer hunters in England and seem to have shown very little interest in hunting them when they immigrated to the New World. There may also have been a moral imperative involved, as stag hunting had been a pastime for the "decadent" upper classes at home in England. Both the Pilgrims and the Puritans would have had no desire to emulate this type of behavior.[350]

While the seventeenth century colonists were inclined to ignore them, a deer population that was "abound" led to other problems that required governmental intervention. A high deer population led to their unwanted incursions into crop and garden fields. *New England's Annoyances,* written in 1647, mentions this problem:

> Even when it is grown to full corn in the ear,
> It's apt to be spoil'd by hog, raccoon and deer.[351]

The death records from King Philip's War include one farmer who grabbed his musket in anger to go out and shoot deer that had gotten into his corn. He was instead killed in the cornfield himself by marauding Natives.[352]

The Massachusetts Court records note the appointment of deer reeves.[353] A reeve in the seventeenth century was appointed to control animal pests, often in response to citizen complaints. These seventeenth-century reeves were not eco-friendly Park Rangers paid to conserve a precious resource. They were hired with the responsibility for keeping destructive wild deer out of lucrative crop fields.

A thriving deer population also created economic problems in what quickly became a cattle based economy. Prices for domestic cattle reached a seventeenth century high within months of the Puritans' 1630 settlements. The already established Plymouth Colony reaped a prolific profit by supplying cows and beef to the newly arrived Puritans.

## SCURVY DÉJÀ VU?

Numerous entries in John Winthrop's journals mention both the high price for cows and the practice of going to Plymouth to get them. Matching the number of animals found in the probate records for post-1630 Plymouth with the few actual archaeological bone assemblages found from the same time period invariably show in-home meat consumption at a much lower rate than the large total number of animals owned. This is indicative of a thriving livestock market where a high percentage of the animals owned were being sold, slaughtered and eaten elsewhere.[354]

Many early frontier towns in the Massachusetts Bay Colony, like Sudbury, were set up on inland meadows specifically as cow towns. The domestic beef market became a key part of the Massachusetts Bay Colony's economy that would later shift focus from the depressed domestic market to provide an important commodity for early trade with the West Indies.[355] In 1692, for example, George Corwin, the sheriff of Salem, was kept busy barreling up meat from confiscated cattle to be shipped at a high profit to the Indies.[356]

The records of the Massachusetts Bay Colony show that economic control was an ongoing governmental concern. The medieval pre-free market economic system called Mercantilism was still in effect. When there was a shortage of wheat in 1641, fearing high prices, the directive was passed that "no baker, ordinary keeper or other person shall bake, to sell or set to sale any bread or cakes made of wheat meale."[357] That same year the building of ships was encouraged because it is "a business of great importance for the common good."[358] Citizens of the colony were encouraged to raise sheep because wool made cloth that was less flammable than cotton.[359] When the high price for livestock brought animals streaming to market from outside Massachusetts, a protective tariff was instituted to discourage competition and protect the farmers of the colony.[360] The supply of meat was protected by regulating the number of swine that could be kept at certain periods of time.[361] In 1643 a moratorium was placed on the killing of calves by butchers, but not those who did so for "their owne use, although they should part with some part thereof to some of their neighbors." As conditions changed this law was rescinded after "some months."[362] By 1685, the slaughter of animals in Boston was such a booming business that there were complaints about "great inconveniences by reason of filth and dirt cast into their slaughter houses and yards of blood and other filth, although such houses and yards are scittuate neare streets and lanes much frequented."[363]

## DISGUISED AS THE DEVIL

The laws concerning deer that were entered into the records of the Massachusetts Bay Colony that are often cited as deer conservation measures were more likely to be related to the economic concerns about this crucial domestic beef market. One, for example, stating that deer could not be sold in Sudbury without governmental approval, was aimed at making sure that venison would not become a marketable commodity that was in any way competitive with domestic meat (especially beef) production. This ban was later expanded to include the entire colony.[364] Hunting bans on deer also seem to be related to, and protective of, fluctuations in the domestic beef market. By eliminating the competition of free venison in the winter months when tracking deer in the snow makes hunting easier, and shipping was limited or affected by the weather, meat prices could be kept fairly stable in the domestic market even if meat supplies were high. In the good shipping months, having some of the domestic meat consumption needs supplied by venison would actually have been advantageous: it allowed producers to reap more profit from the barreled beef that could be sent to the West Indies.

Another factor related to the deer population was the fact that at first, even if the settlers wanted to hunt for deer, it would have been difficult. The settlers' earliest weapons, especially the matchlock musket, were not conducive to either the pursuit or shooting of deer. They were awkward and difficult to aim. They were so heavy that they needed a stick stand to hold them up while firing. The firing mechanism was dependent upon having a lit wick (match) at the time of use. The awkwardness of this is evident in the events that occurred during the first two expeditionary forays of the Pilgrims on Cape Cod.

Sensing that they might be near a Native American village, these early explorers had to stop and make a fire to light their matches (actually a sort of smoldering wick) so they could be prepared to fire their muskets. Any deer (or Native Americans) that had not already run away would have been frightened away by the smell of smoke and fire.[365] Later, during their "first encounter" with attacking Natives, part of the battle action consisted of having to pass around a burning log to light matches before the muskets could be shot. Once shot, the heavy muskets were dropped and replaced by cutlasses and sabers when the Pilgrims charged after their, by then, fleeing attackers.[366] Since deer can run faster than humans, especially those who are weighted down with chain mail and armor, the Pilgrims' sabers and cutlasses were also probably not good deer hunting weapons.

## SCURVY DÉJÀ VU?

In that 1620 encounter only one man had a snapchance, the earliest musket that used a spark from a flint to ignite a shot. Snapchances would become rapidly popular. By 1630, the Massachusetts Bay colonists had 300 of them on their inventory list to bring with them to the New World. However, it took until late in the seventeen century for the flintlock musket to be perfected, and until between 1720 and 1740 for the barrel to be rifled (have grooves inside that put a spin on a shot as it is fired). This created a light, accurate, aimable, easily fired weapon-- the American long-rifle.[367] It was only in the early eighteenth century, as New England's culture matured and began to deviate from the established patterns that had been initially brought from England, that this rifle was aimed at deer. It was the eighteenth century that would see a dramatic decline in New England's deer population.

The focus of the English settlers in Massachusetts at first was on fishing, building homes and farms, and on their domesticated animals, not deer. It is difficult to analyze seventeenth-century deer usage in Plymouth, other than the occasional written mention of deer hunting, because the very early attempts at archaeology tended to throw out animal bones and remains as unimportant. The few modern archaeological studies that have been done have turned up a very low number of deer bones in the assemblages.[368] In Boston, however, it can be stated that they were not eating much venison in the seventeenth century. The Boston Cross Street privy archaeological site 4 shows that deer were an extremely rare element in the colonists' diet at any time during the mid to late seventeenth century. [369]

But the forest fringe environment around English settlement was in for a dramatic change. The abundant deer and lots of other species were going to be driven away from settled areas by the behaviors of one of the most prolific and environment altering of the new residents-- the pig.

 DISGUISED AS THE DEVIL

# THE ENGLISH COLONISTS IN MASSACHUSETTS WERE
# BEEFEATERS, NOT DEER HUNTERS

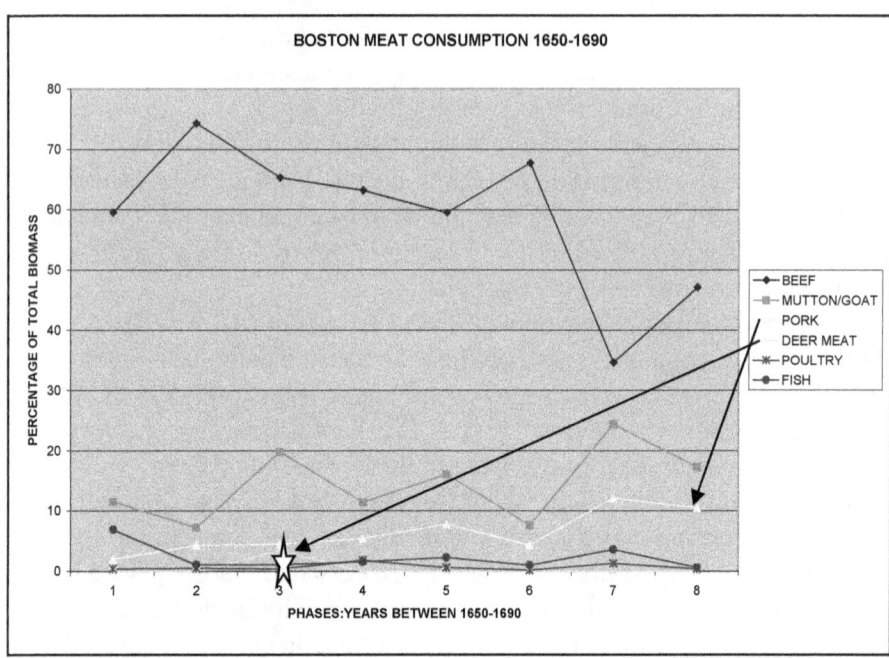

**Figure 29.** Level 1 is at +/-1650 and level 8 at +/- 1690. It shows that one early Boston household clearly preferred beef and was not eating any venison (deer meat) except for a short time in the 1670's when it was at 3.9 percent of the total (possibly during the King Philip's War period) as depicted in PHASE 3 on this graph. Consumption of fish was also low. This consumption pattern may be typical for early Massachusetts showing that there was probably not much deer hunting going on in that century.

From Gregory Brown and Joanne Bowen, "Animal Bones from the Cross Street Back Lot Privy" in Ronald Michael, Ed. *Historical Archaeology* 32 No. 3(1998).

# SCURVY DÉJÀ VU?

# SCURVY DÉJÀ VU?

330 John Pond letter in Everett Emerson, Ed., *Letters from Massachusetts Bay Colony 1629-1638* (Amherst: University of Massachusetts Press, 1976).
331 Willison, *Saints,* 270.
332 Interview with Dr. Sabra Bellovin, WABC, November 7,2007, "The Lyme Controversy" at www.wset.com/news/stories/1107/470683.html
333 Janick, J. " Herbals: The connection between horticulture and medicine," *Hort Technology* (2002), See also Sherman and Billings, *Darwinian Gastronomy.* Lemon juice also acts to enhance the properties of other spices, making them more potently antimicrobial in combination.
334 Willison, *Saints,* 270-272.
335 Eric Christianson, *Sickness and Health in America* (Madison: University of Wisconsin Press, 1997),48. See also Eric Skaar, "Did Bloodletting have benefits?," *Science* (September, 2004) for a recent study that found that bleeding may actually be beneficial to patients with a staphylococcus infection because it starves the bacteria of iron.
336 John Winthrop, Jr., unpublished letter from *Letters of John Winthrop, Jr.,* Massachusetts Historical Society Collections.
337 John Winthrop was opposed to this move to Connecticut and has extensive entries in his Journal in the 1630's discussing the topic. The legislation allowing the move begins in *The Records of the Colony of Massachusetts Bay in New England,* Volume 1, on page148.
338 John Winthrop, Journal, 200.
339 Winthrop, *Journal*, November 17, 1636 entry.
340 An autopsy done during the seventeenth century also mentioned this burned look of some of the tissues of a man deceased from witch affliction-(Philip Smith,1684) in Cotton Mather's, *Magnalia Christi American,* (Cheapside, London: Thomas Parkhurst, 1702).
341 Hazel Adele Pulling, "California's Range Cattle Industry, Decimation of the Herds, 1870-1912," *Times Gone By, The Journal of San Diego History,* Volume 11, Number 1(January, 1965).
342 David Cressy, *Coming Over* (Cambridge U.K.: Cambridge University Press, 1987), 280-281.
343 Glenn Motzkin and David Foster, "Grasslands, heathlands and shrublands in coastal New England: historical interpretations and approaches to conservation," *Journal of Biogeography* 29(2000), 1569-1590.
346 Winthrop, Journal,200.
345 . See Appendix H. Model derived from information in accounts of Higginson, Winthrop, Bradford and Winslow and from several figures in D.R. Foster, et al., "Wildlife dynamics in a changing landscape" *Journal of Biogeography* (2002): forest, deer and wolves 1339, deer 1344, Tom Wessel's *Reading the Forested Landscape: A Natural History of New England* ideas about forest levels, and Daniel Gookin's estimates of Native American population levels .
346 Woods, *New England's Prospect,* 59. See also John Winthrop, *Journal,* December 26, 1631 entry.
347 Nathanial Shurtleff, *Records of the Governor and Massachusetts Bay in New England,* 232.
348 James Duncan Philips, *Salem in the Seventeenth Century* (New York: Houghton Miflin, 1933), 36.
349 G. Motzkin, D.R. Foster, Debra Bernardos and James Cardoza "Wildlife dynamics in a changing landscape," *Journal of Biogeography* (2002), 1337-1357.
350 Charles Cheek, "Massachusetts Bay Foodways: Regional and Class Influences," *Historical Archeology* 32 No. 3(1998), 156-159.
351 J.A. Leo Lemay, *New England's Annoyances: America's First Folk Song* (Newark, Delaware: Associated University Presses, 1985), Prologue.
352 Philips, Salem, 36.
353 John Winthrop mentions deer reeves in his writings. A reeve is defined as a bailiff or a minor officer of parishes or other local authorities in Mark Boyer, ed., *The American Heritage Dictionary* (Boston: Houghton Mifflin, 1991), 1038. The theory that deer were a scarce commodity that reeves and hunting regulations sought to preserve appears first in William Cronan's *Changes in the Land* and is then repeated in other works.
354 See Livestock of Plymouth colony 1620-1692 Research Report by Craig S. Chartier at Plymouth Archaeology Rediscovery website http://plymoutharch.tripod.com for an extensive analysis of archeological sites and 257 probate records.
355 David Cressy, *Coming Over,* 280.
356 Ibid., 279.
357 Nathanial Shurtleff, *Records,* 337.
358 Ibid.
359 Ibid., 232.
360 Ibid., 293, repealed later, 309.
361 Ibid., 148.

362 Ibid.37.
363 Ibid., Section 4, 37.
364 Ibid.,208
365 Bradford, *Plymouth*, a description of this first encounter is on 76-77.
366 Winslow, *Mourt's*, 36.
367 Kenneth Chase, *Firearms: A Global History to 1700* (New York: Cambridge University Press, 2003), 23-26 and Shurtleff, Records, *inventory list 1630*.
368 James Deetz and Patricia Scott Deetz, *The Times of Their Lives* (New York: Anchor Books, 2000), 246-247.
369 Gregory J. Brown and Joanne Bowen, "Animal Bones from the Cross Street Back Lot Privy," Ronald L. Michael, ed., *Historical Archeology* 32 No.3 (1998).

THE PIG

## VI. THE PIG IN THE PROMISED LAND:
The witch "made a noyse and turned into the shape of a blak **hoge**…"
Joseph Ring, Salem, Massachusetts, May 13, 1692.[370]

The pig species, *sus scrofa*, played a role in the etiology of witchcraft and an important role in the European settlement of the New World. The Vikings had swine with them for the initial phase of colonization in Greenland.[371] The English settlers seem to have followed the same practice. Pigs were an environment altering force in the forests of New England. It is not recorded when pigs were first set loose in Plymouth but a few of them may have been passengers on the Mayflower.

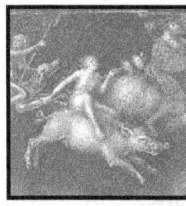

**Figure 30**. *A witch riding a boar-they are often connected because both inhabited the forest edges, from* Melancholy *by Lucas Cranach, the Elder, 1530.*

Edward Winslow specifically leaves swine and chickens out of the 1621 list of animals needed by Plymouth when he wrote, "if we have once but kine, horses and sheep, I make no question, but men might live as contented here as any part of the world." William Bradford noted the arrival of cows "the first of that kind of cattle [perhaps implying that other kinds were already there- goats, pigs and even chickens]"[372] in Plymouth. By 1624, there were "many swine."[373] The Massachusetts Bay Colony is also described as having swine aplenty. In 1629 Francis Higginson wrote from Salem that "it is scarce to be believed how kine and goats, horses and hoggs do prosper here."[374]

One early Puritan settler wrote in 1631 that "the best cattle for profit is swine."[375] First person accounts of early settlement mention there being "many swine" and there are extensive accounts of pig behaviors that have become problems. There are records of votes at town meeting to hire at public expense one or two keepers for town herds.[376] The English

## DISGUISED AS THE DEVIL

medieval pig husbandry practice of a keeper watching the town's hogs seems to have prevailed in some areas, but there still seem to be a lot of pigs running around, based on the number of complaints mentioned in the Massachusetts General Court records. Swine are noted to be allowed to range free in the woods from November to March. [377] Some towns kept records of pig earmarks. Each family would have a distinctive notch that was cut into the ears of the pigs that they owned so that they could be identified by their owners.[378]

This onslaught of prolific swine had to have made an impact on the ecology of the newly settled areas of Massachusetts. What did they do when they were turned loose in the new world? Like their human owners, seventeenth-century pigs seem familiar because of their modern descendants, but may have actually been as different from them as we are from our seventeenth century ancestors.

What happens when pigs run free? Porcine characteristics and behaviors that might have been both understood and exploited by seventeenth-century pig owners are difficult to find in the modern domesticated pig's caged world. However, by looking at those few places where pigs now live freer lives, some important behaviors become evident. The twenty-first century world has not caged every pig. There are still areas where pigs roam free. It was possible to find a modern body of information about these pigs and study it for characteristics and behaviors that might be useful when applied to the past. One of the best sources of information is the experience of those who have had to deal with feral pig populations.

Modern California has a pig problem. Various inhabitants including park rangers, wildlife control officers, environmentalists, golf course supervisors and the occasional Buddhist monk are battling this "wily" pig foe. That state's ever enlarging feral pig population was pegged at 135,000 and rising in 1999. At Pinnacles National Monument in San Benito County, for example, a walk through any part of the 24,000-acre park was described as similar to what one would experience when "an army of roto-tillers had run amok." This is the work of rooting pigs; the park has built a thirty-mile long fence at a cost of $40,000 per mile to get rid of them. [379]

California's wild pigs are the direct descendents of a set that was intentionally let loose in the region in 1925 for sport hunting. Once loose, they became enamored with a few escaped domestic pigs and founded a genetically mixed herd. Even before the interest in pig hunting waned, this population was well established and since then it has skyrocketed.

# THE PIG

When hunting was re-introduced as a control mechanism in the late twentieth century, it was found that these pigs were more than a match for even the fiercest of hunting dogs.[380]

In California, the pigs have adversely affected many native species, including deer, squirrels, quail and other birds, by out-competing with them for food, especially acorns. Joe DiDonato, a wildlife program manager for California's East Bay Regional Parks, notes, "we have been looking at the stomach contents of some of the pigs that we have killed and we find them stuffed with acorns. When we go out looking for acorns in the wild we cannot find any at all on the ground." This sentiment is echoed by Cody Stember, a professional trapper with the U.S. Department of Agriculture, who wrote, "they're out there right now sucking up acorns like a vacuum. They'll run the deer away."[381]
A recent study of the home range movements and habitat of these wild swine found that pigs will root anywhere but tend to root the most under oak trees. They tend to have a "home" territory which is not overly large. In California, this rooting behavior is called "pig plowing." These pigs prefer dense brush or marshy vegetation. Because they don't have a sweat mechanism for body cooling, during periods of hot weather pigs spend a good deal of time wallowing in ponds, springs or streams, usually in or adjacent to cover (the Buddhist monks awoke one morning to find a tranquil reflective meditation pool with a carefully [and costly] contrived landscaping of imported flora had been transformed into a mud filled wallow.) Pigs will eat anything from grain to carrion. There are anecdotal reports that pigs will attack and eat unattended fawns.
From this information some behaviors that may have been important in seventeenth-century pigs can be delineated. The preferred food for the pig is the acorn, which was usually abundant in early Massachusetts. Pigs are territorial and their territories tend to be concentric rings around oak trees or other food sources. In a given geographic area, they will out-compete native species for food resources, most notably deer and birds. They can handle the attack of a fierce hunting dog (roughly analogous to a wolf). Their rooting behavior, called pig plowing in California, can rapidly denude an area of vegetation bringing the ground down to bare soil. When this behavior, useful at first to clear land, became undesirable, it was dealt with by the seventeenth-century practice of putting a ring through pigs' noses.
While pigs were not a favorite subject for artists, some depictions from early time periods exist both in Europe and in the United States. Based on those depictions, seventeenth-century pigs seem phenotypically similar

## DISGUISED AS THE DEVIL

to the feral mixed breed pigs in California. Described as "little brown striped watermelons" that squeal, California's new born piglets may be living reflections of the earliest *sus scrofa* inhabitants of Massachusetts.[382] In the *Good Cheape Husbandries,* written in 1614 by Gervase Markham, he suggests that the best British swine should have a "thicke neck, a short and strong groyne, and a thick chine well set with strong bristles. The colour is best which is all of one peece, as all white or all sanded, the pyed are the worst and most apt to take the mesles, the black is tolerable." Pig disposition is described as being "greedy, given much to root up grounds, and teare down fences, he is very lecherous. he is subject to much anger." Markham states that in England swine were fed in the morning and then brought out by either families or a hogmaster to root in either old fields, marshes, or (in the fall and winter) to the forests for mast during the day and then returned to the safety of the sty at night.[383]
A study of both modern and historic farming practices in England reveals another important fact related to the pig's preference for acorns. In *500 Points of Husbandrie*, a poem written by Thomas Tusser in 1577, the following lines can be found:

> To gather some mast, it shall stand thee upon
> With servant and children, yer mast be all gone:
> Some left amongst bushes, shall pleasure thy swine,
> For fear of mischief, keep acorns from kine. [384]

A mature Oak tree can produce up to 700 pounds of acorns in a good year.[385] Gathering mast required all available hands to work at that time including those of servant and child. The verse suggests leaving some acorns on the ground for swine. It adds the important fact to keep acorns from cows. The pig and the goat, another animal abundantly represented in very early seventeenth-century Massachusetts, can digest acorns. However, when they are eaten by either cows or horses, acorn poisoning can cause sickness and death.[386] Knowledge of this fact would have been important in 1692 because English settlement had reached to the edge of the wilderness in the "Frontier" town of Andover and in Salem Village. Cramped for pasturage, both horses and cows were being kept "in the woods."[387] Knowledge of this fact is at the root of the modern English practice of "pannage." Using an archaic sounding set of calendar dates straight out of the medieval mind, this period traditionally lasts from September 29 to fifteen days before Easter hocktide, but may be changed to allow for weather fluctuations. During this time, English farmers may assert their "right to mast." Pigs are allowed onto open

# THE PIG

forestland to eat acorns and, by so doing, keep cows, ponies and horses that graze there in warmer weather free from acorn poisoning.[388]

The "pig plowing" that is happening in California today shows that swine are a good land clearing animal. Centuries of pig domestication in England made Massachusetts's English settlers well

Figure 31. *A 17th C. depiction of pigs: English woodcut print of "the swimming of Mary Sutton" from <u>Witches Apprehended, Examined and Executed for notable villainies by them committed both by land and water. With a strange and most true trial how to know whether a woman be a witch or not.</u> London, 1613. Note that the sows depicted have a "wild" phenotype.*

## DISGUISED AS THE DEVIL

Figure 32. *14th C. depiction of pigs, with their keeper, eating acorns during pannage. Note that they also display wild characteristics. "November" from <u>Hours of Jeanne d'Evreux</u> by Jean Pucelle 1325-28.*

aware of this pig habit. Plowing pigs could clear land without requiring any intensive human effort. While one was occasionally mentioned as being killed by a wolf,[389] it is an infrequent notation. By using the knowledge of their pigs' territorial behavior, the settlers in early Massachusetts could let their pigs inhabit the forest edges with little concern that they would run away. The territorial sense of "home" could be reinforced by the regular presentation of food and garbage. The settlers had little to worry about maintaining their pig's food, as acorns (mast) were readily available in the Massachusetts forest that was consistently described as being predominated by oak trees. Archaeological evidence, in the form of pollen core sampling, supports the preponderance of oak trees in Massachusetts seventeenth century forests.[390]

Because of their territoriality, in years when acorn production was diminished, the local swine became garden pests, exploiting weak areas in their owners' fencing. John Winthrop wrote in September of 1633, "There was great scarcity of corn by reason of the spoyle our hoggs had made at harvest and the great quantity they had eaten in the winter (there being no acorns.)"[391] This seems to support both the need for strong

## THE PIG

fences to keep out pigs, laws requiring pigs to be "yoked," the territoriality of the local swine, and the possible collection, as recommended by Thomas Tusser, of acorns for winter-feeding. The "gathering of akrons" is mentioned in a November 5, 1639 Massachusetts General Court order regulating outdoor 'fyers.'[392] A lack of acorns led to the substitution of corn.

Following standard Middle Age practice, outside gardens were fenced. By 1624, Plymouth, had "gardens encased in clapboards" and "the towne" according to John Smith "is impaled."[393] Swine, wolves, French ne'er-do-wells and hostile natives were kept at bay in this controlled manner. This was a lingering medieval farming strategy. Early English archaeological sites often show a round fence, ditch, or dirt wall thrown around an entire site for both defense and animal control. The initial choice of terrain, especially in the case of both Plymouth and Boston, corresponds to the ancient English hill fort settlement pattern.[394]

Seventeenth-century swine husbandry practices had the direct result of creating an expanding Pig Rooted Zone (hereafter PRZ) of ground around any site occupied by their human owners. By regularly offering food to their pigs in a certain area owners insured that a point became the center of the animal's home territory. If the food that was offered included collected acorns, a farm or town itself might have been synonymous to a well producing oak tree in the seventeenth-century pig's mental landscape.

John Winthrop, Jr., in a 1664 letter to his friend Lion Gardner of Long Island, recommended this pig plowing mechanism as having been successfully used in the Massachusetts Bay Colony. He wrote that pigs and goats "closely grub the earth," which once cleared in this manner could then be planted with "proper English grass to create a meadow for cattle, horses or sheep."[395]

Similar to the impacts observed in California, an immediate effect of this PRZ would have been the diminishment of a preferred food source for at least five species indigenous to seventeenth-century Massachusetts: the now extinct migratory passenger pigeons, wild turkeys, white footed mice, and white tailed deer.[396] *Ixodes scapularis* tick populations would have been affected by the constant stirring and rapid decomposition of leaf litter caused by rooting pigs. While mice, deer, and ticks seem to have had co-existed in the oak forests of New England for eons, co-existence with pigs would have been at best difficult. A mouse with any sense whatsoever would not take on a large sharp-toothed pig to get an acorn. Even a deer would have stepped aside for these nasty new acorn consumers. Since wild pigs have been observed to kill and devour fawns in California, this behavior may also have had some effect on a local deer population. As is happening in California, mice, birds, and deer would have been driven away. To survive, indigenous species would have retreated outside the PRZ. This competition may have been partially responsible for William Bradford's statement that wild fowl "is not as plentiful hereabouts as it once was."[397]

The now extinct but once prolific passenger pigeons so numerous that that they blackened the skies as they flew over, ate acorns. On September 26, 1633, William Hammond of Massachusetts Bay Colony wrote, "when it is an acorn year here our pigeons abound."[398] Pig related acorn depletion may have been the first nail in that specie's coffin of extinction, followed by an apparent seventeenth-century consumption level that was equal to the domesticated chicken, at least in Boston.[399] In California, a distant relative, the Bandtail Pigeon, still gorges on acorns before migrating south to Mexico for the winter and is considered threatened by pig acorn depletion.[400]

Acorns are very high in fat and carbohydrate content. Birds, mice, deer and pigs that ingest large quantities of acorns in the fall gain weight and bulk up for the winter ahead. Studies have shown that deer can gain weight rapidly in just two weeks of this acorn gorging. Birds gain strength, body mass and energy for their long migratory flights.[401]

Humans in early New England exploited this acorn driven pre-winter weight gain in their pigs by turning late Fall into slaughter time, usurping the added body mass for their own dinner tables. This custom follows the English November through Shrovetide (late February) slaughtering pattern.[402]

New England's tick populations are sensitive to any movement, growth or degradation in the white tailed deer population. Pig driven movement of deer would have also moved ticks. When the deer around English settlements were crowded out by the resident pigs, any attached deer ticks would simply have moved along with them. Those left behind would not have flourished. A leaf litter layer in a PRZ was not a protective feature, especially in cold weather. It was constantly being churned up. Tick eggs would have been susceptible to being exposed to below freezing temperatures if they were churned to the surface during the winter. After the initial phase of sickness at contact, records show that the English populations in early Plymouth and Massachusetts became relatively healthy. While human health became fairly robust, Massachusetts's pigs were not so lucky. Pigs were constantly "at risk" for

## THE PIG

Lyme disease because they occupied almost exclusively the geography of risk for Lyme disease--between the forest and cleared land.

This underscores a critically important function for the PRZ as a biological sanitizer. Pigs on the perimeter of human occupation zones were in a position to carry the brunt of possible Lyme infection. They would live a lifecycle shortened by annual slaughter, which would have the effect of limiting the ongoing spread of tick borne disease. If the pork from an infected pig was cooked to a temperature above 160 degrees F. the Lyme bacteria was killed, which ended any further transmission of the disease.[403]

This PRZ function may be corroborated by the bones of some of the earliest pigs that have been found in Massachusetts. During archeological work done in association with Boston's "Big Dig," 371 bones representing the heads, bodies and feet of seventeenth century pigs were found, as well as a nearly complete fetal pig carcass. One of the most important findings in these pig remains was evidence of fairly widespread pathologies. One bone, a scapula, contains massive exostoses (pitting from infection) around the periphery of the glenoid (socket); there also appears to be an ankylosis (joint fusion) evidenced by a rough and pitted surface on the glenoid that has extended beyond the original surface. The most interesting pathology is from an immature pig skeleton. This appears to be a fetus that would have been described as being "cast off" or aborted in the seventeenth century. The skeleton had a general swelling on the distal half of a tibia, which may have been the result of an infection of the bone via an infection in the blood stream. Lodging in the Haversian system (bone marrow and other parts of the bone), especially in a young animal, such infections can cause death.

While pigs are known to suffer from a variety of infectious diseases, all of these pathologies are consistent with a Lyme disease diagnosis for these pigs.[404] If so, they indicate that these animals (including the fetal pig's mother) were active participants in the PRZ process. By occupying the edges of the English cultural zones, seventeenth-century pigs may have helped also carried the brunt of Lyme disease infection and kept humans healthy.[405]

In seventeenth-century Massachusetts, the English and their domesticated animals exploited land that had been cleared first by Native Americans, and then by pigs, of brush, undergrowth, acorns, deer, mice, innumerable insects whose life cycle included an underground grub stage, and ticks. An ongoing cycle would be repeated as a new generation of pigs ventured further a field in an enlarging zone of pig and human occupation. When

## DISGUISED AS THE DEVIL

combined with the English settler's continuation of the medieval custom of "burning" fallen leaves and undergrowth, the bio-sanitizing system seems to have worked very well.

The Native American residents of Massachusetts' version of the PRZ was their collection of acorns to use as food and their protective slash and burn agricultural system and burning of undergrowth. This ridded the area where they lived of not only weeds but acorn eating mice and insect vermin as well. This system, if repeated regularly, also prevented endemic Lyme disease from ever reaching epidemic proportions.[406] This healthy cycle began to break down first, when the Native American population was decimated, then English settlement began to fragment the forests, and when once desirable pig behaviors became a nuisance that had to be controlled. This happened in stages. There are innumerable seventeenth century Massachusetts Bay Colony ordinances related to swine. The Massachusetts General Court tried many strategies to respond to an onslaught of complaints. They told complaining Native Americans to build fences. They tried fining the owners of wayward pigs. They tried fining people who complained about the Hog Laws. They appointed "Hogg reeves" to chase after pigs. They banned fish stealing pigs from the beaches near fishing platforms. They tried impounding pigs. They allowed pigs found in cornfields to be shot. They tried nose rings and yokes that enlarged the neck size of the pig to make it harder for them to exploit the weak areas of fences. By the late seventeenth century, rings and yokes were legally mandated in most areas.

Gradually, and later by legal mandate, domestic swine were relegated to the fenced farmyard. By the 1690's pigs are mentioned as sometimes being confined within fences. Pregnant sows seem to have given birth within the confines of their owner's barnyard and owners were well aware of the number of offspring and the health of their animals. One seventeenth century witchcraft deponent verifies that pigs have begun to be caged when he testifies that his bewitchment consisted of being thrown into a neighbor's sty.[407]

Once the pigs were brought in from the forest for good, however, Lyme disease was still being controlled by some prevailing cultural practices: a new interest in hunting deer, acorn collection as hog food, the selective extraction of oak trees, and the burning of fallen leaves and undergrowth. Most people in these areas lived an outdoor agrarian lifestyle. However, by the mid twentieth century, New England's farming had declined and many previously cleared fields once again underwent the natural process of reforestation.

## THE PIG

During the nineteenth century, acorns were replaced by corn as the major feed for hogs. This left the next century's forest's floor littered with mast for birds, mice, and especially deer. The decline in and regulation of recreational hunting led to a concurrent dramatic increase in the number of white tailed deer in those reforested woods.[408] In addition, in 1970, the Federal Clean Air Act suggested that states ban the yearly burning of fallen leaves to preserve air quality.[409] This removed the last existing control for tick populations. It immediately and dramatically expanded the over-wintering habitat for *Ixodes scapularis* ticks. Portions of New England's burgeoning suburban population were poised on the brink of a new epidemic of Lyme disease that, if not for the availability of antibiotics, might have rivaled the "afflictions" of the seventeenth century.

## THE PIG IN THE PROMISED LAND

[370] Essex County Archives, *Salem Witchcraft* Vol.1, 64.
[371] Fitzhugh and Ward, *Vikings*, 331.

[372] Cows were called kine, William Wood, *New England's Prospect*, 57, and Bradford, *Plymouth*.
[373] John Smith, *General Historie of Virginia, New England and the Summer Isles*, (London, 1624) Reprinted in Philip Barbour, ed., *Complete Works of Captain John Smith*, Volume 2 (University of North Carolina Press, Chapel Hill. 1986), 472-473.
[374] Francis Higginson letter in Everett Emerson, Ed., *Letters from Massachusetts Bay Colony 1629-1638* (University of Massachusetts, Amherst 1976).
[375] *Ibid.*, Thomas Graves letter.
[376] Virginia D. Anderson, *Creatures of Empire* (New York: Oxford University Press, 2004), 161-162.
[377] Numerous cases, laws, and complaints are located in the Nathaniel Shurtleff, *Records of the Governor and Company of the Massachusetts Bay in New England* (AMS Press, New York, 1968) see 87,101,104,106,110,119,148-50,157,181,187-89,215,238-39,255,219,222,265 and 270.
[378] Earmark records are often found in unpublished town papers. Earmarks are discussed on page 143 of *Historical Sketches of Andover* (Boston: 1880) also see *Earmark Record Book* of Oyster Bay, Long Island for seventeenth century earmarks.
[379] *California's Feral Pig Problem* from" Onslaught of the wild pigs/Bay area Park invaders march northward despite trapping efforts" at www.sfgate.com
[380] California Department of Environmental Science Policy and Management, *Wild Pigs* (Berkeley, California: California Printing Office, 1998).
[381] www.sfgate, *California's Feral Pig*.
[382] F.J. Singer, D.K. Otto, A.R. Tipton, C.P. Hable, "Home range movements and habitat use of Wild Boar," *Journal of Wildlife Management 5* (1981), 263-270 and J.J. Meyer and I.L. Bribin, Jr., *Wild Pigs of the United States-their history, morphology, and current status* (Athens, Georgia: University Press, 1991), 313.
[383] Gervase Markham, *Good Cheape Husbandries* (London Reprint of 1614 edition).
[384] Thomas Tusser, *Five Hundred Points of Good Husbandrie* (London, 1577, Reprint 1878), 32.
[385] Paul Johnson, S. Shifley, and R. Rogers, *The Ecology and Silvaculture of Oaks* (New York: CABI Publishing, 2002), 64-67, 77.
[386] www.hants.gov.uk discusses acorn poisoning in modern times.
[387] See Salem Witchcraft testimony of William Barker, Jr., *Suffolk Court Records*, Case No. 2761, 102.
[388] www.hants.gov.uk also has a discussion of pannage.
[389] John Winthrop, *Journal*, entry of August 20, 1631.
[390] Gerald Kelso, "Pollen Analysis of the Cross Street Back Lot privy," in Ronald L. Michael, Ed., *Historical Archeology* 32 No.3 (1998), 53.
[391] John Winthrop, *Journal*, September, 1633.
[392] Shurtleff, *Records*, 90.
[393] John Smith, *Historie*, 472.
[394] Conrad Cairns, *Medieval Castles* (U.K.: Cambridge University Press, 1987), 6-8 for a good discussion of the siting of hill-fort strongholds.
[395] John Winthrop, Jr., from *Letters of John Winthrop, Jr.*, Massachusetts Historical Society Collections.
[396] See Peter Alden and Brian Cassie, *National Audubon Society Field Guide to New England* (New York: Knopf, 1998) for a list of extant wildlife. Passenger pigeons are mentioned in Paul J. Lindholdt, ed., *John Josselyn, Colonial Traveler: A Critical Edition of Two Voyages to New England* (Hanover and London, 1988), 71.
[397] Bradford, *Plymouth, op. cit.*
[398] William Hammond in *General Considerations for Planting New England: Chronicles of the colony of Massachusetts Bay 1623-1636*, September 26, 1633
[399] Brown and Bowen, "Animal Bones from the Cross Street Back lot privy," in Ronald L. Michael, Ed., *Historical Archeology* 32 No.3 (1998), 75.
[400] www.sfgate.com.
[401] Johnson, *Oaks*, 81-84, also see acorns at www.bowsite.com.
[402] See page 7-8 of *Livestock of Plymouth colony 1620-1692 research* report by Craig S. Chartier, Plymouth Archaeology Rediscovery website at http://plymoutharch.tripod.com.
[403] Karen Vanderhoof-Forschner, *Everything*, 155.
[404] My experiences with modern pigs raised in a historic farm setting in a Lyme disease endemic area is that they can show similar joint infections.

[405] Brown and Bowen, "Animal Bones from the Cross Street Back lot privy," in Ronald L. Michael, ed., *Historical Archeology* 32 No.3 (1998), 77. Evidence of disease in pig remains included a trephined skull, exostoses on a scapula, and swelling on shaft of tibia.
[406] M.L. Wilson, "Reduced abundance of adult *Ixodes dammini* following destruction of vegetation," *Journal*

# THE PIG

*of Economic Entomology* 79(1986), 693-696.
[407] George Lincoln Burr, *Narratives of the Witchcraft Cases (*Mineola: Dover Publishing, 2002 reprint of 1914 edition), 140.
[408] David Foster, et al., "Wildlife Dynamics in the changing New England Landscape," *Journal of Biogeography* (2002), 1338-1339.
[409] *The Federal Clean Air Act,* 1970.United States Code Title 42-Chapter 85.

# DISGUISED AS THE DEVIL

**Why Women?** Almost all of the women that arrived in 1620 aboard the Mayflower died. Accused witches were also overwhelmingly female and in Salem Village, the majority of the afflicted were also female. Could the way women dressed be partially to blame?

**DRESSING THE PART-AN EXPERIMENT**

Experimental design: Two modern humans would collect kindling for 30 minutes in an oak/pine wooded area on a warm October day in an area know to be endemic for Lyme disease. A male and a female would wear a white biohazard suit under a set of other clothing. The woman would wear a historic costume from the seventeenth century, which included woolen stockings, a woolen outer skirt, a linen inner petticoat, a cotton chemise and an apron. The man would wear woolen pants that he tucked into long woolen stockings. After participating in the activity all clothing would be stripped off and *Ixodes* ticks would be collected from each set of clothing and counted.

Results: Male Subject: 14 ticks
       Female Subject: 56 ticks

Discussion: Modern precautionary directives may be effective tick controls. In a historic context, women would be at a disproportionate risk for Lyme disease because of the style of clothing worn: **the long skirt**.

# THE WITCH

## VII. The Witch at the Edge of the Woods?

If any man or woman be a witch, that is, hath or consulteth with a familiar spirit, they shall be put to death. [410] *Statutes*, Massachusetts Bay Colony, 1641

"One sort of such as are said to be witches are women which be commonly old, lame, blear-eyed, pale, foul, and full of wrinkles; poor, sullen, superstitious…in whose drowsy minds the Devil hath gotten a fine fear so as, what mischief, mischance, calamity, or slaughter is brought to pass, they are easily persuaded the same is done by themselves…" Reginald Scott 1595[411]

**Figure 33**. *This depiction of a witch was printed by John Hammond in 1643 on the cover of <u>A Most Certain Strange and True Discovery of a Witch.</u> She has distinct features that include being old and wrinkled, using a staff to walk, and having bird familiars along with what look like Semitic prayer strings tied to her waist.*

In *The Disenchantment of Magic: Spells, Charms, and Superstition in Early European Witchcraft Literature,* Michael D. Bailey writes that one of the hallmark features of modern western thought, born of the protestant reformation but which came into full vigor during the Enlightenment, is what Max Weber called "the disenchantment of the world." This entails belief in the primacy of science and that "there are no mysterious incalculable forces" in the world. Issues of "magical thought" and "superstition" in opposition to "scientific rationalism" frame many discussions in which western modernity is juxtaposed against the traditional beliefs and practices of the historic past. [412]

In the case of witchcraft, the disenchantment of the world caused the proverbial baby, in this case eyewitness observations of red marks on

## DISGUISED AS THE DEVIL

the skin of witches and the afflicted, to be thrown out with the bathwater. Surely many of the ideas related to witchcraft have been rightly determined to be superstitious, the products of the overzealous imposition [often accompanied by the use of physical torture] of Christian dogma on the populous, or the scapegoating of women- but there may be an enlightened or even scientific explanation for some of the events, folkloric beliefs and practices and there may be some veracity in physical observations made in the past.

Once again, the ecological history of the landscape involved plays a crucial role as the stage on which the story of the witch is played. In Europe, where the beliefs in witchcraft that were carried to the New World originated and evolved, there were cyclical waves of human population expansion and decline and deforestation and reforestation. Human activity, especially the adoption of a sedentary farming lifestyle, altered the forested landscape when trees were cleared to build houses and make way for crop fields. The concentration of grazing animals- cattle, sheep, goats and especially pigs, in the areas around human settlements altered the landscape in other equally dramatic ways. As populations moved, expanded or contracted this modified environment also changed. It contracted after population decline and was followed by reforestation, especially after the wave of plague called the Black Death.

During the middle ages, European population levels varied over time. Starting at a post Justinian's Plague low point of 542 AD, there were considerable fluctuations that accompanied both growth and decline. The populations of early medieval settlements tended to be small and clustered together with large zones of forest in between. Populations rebounded slowly after disasters, and disasters could and did strike rapidly with devastating results. After the population depletion of the sixth century, it took until the eleventh century for the population to grow large enough for the "great clearances," when many of Europe's forests were cut down, to begin. The European population expanded beyond the Elbe River into Eastern Europe and the crusaders interacted with the Middle East. This time period was accompanied by a warming climate known as the Medieval Warm Period.

By about 1300, the population of Europe had reached a peak of between 70 and 100 million before calamities struck again. The Great Famine of 1315 was followed by the Black Death or Bubonic Plague and, not surprisingly, The Hundred Year War that actually lasted for the 116 years between 1337 and 1453. It is estimated that the population of Europe decreased by as much as forty percent between 1348 and 1420. [413]

## THE WITCH

Dead people do not cultivate fields. This is clearly evident in the pollen record. This period of population decline was followed by a period of brushy re-growth and reforestation, and eventually, as populations recovered in the early sixteenth century and then increased, another wave of deforestation.[414] The farm abandonment that accompanied the nineteenth century's Industrial Revolution also led to reforestation at a later time period, which was then followed by the deforestation of the twentieth century's modern suburban development. Lyme disease symptoms would be most likely to appear during any of the "contact" time periods and be influenced, as we have seen, by moisture, masting, and deer and mouse population levels as well as cultural practices.

The history of the European concept of witchcraft can be interwoven with this environmental history. A deeply rooted and prevailing folk or pre-Christian belief system included ideas about the supernatural. As Rome declined and Christianity was introduced, it incorporated witchcraft into a prevailing belief set with the Old Testament admonitions of Exodus 22:18 that: "Thou shalt not suffer a witch to live" and Leviticus 20:27's: "A man also or woman that hath a familiar spirit, or that is a wizard, shall surely be put to death: they shall stone them with stones: their blood shall be upon them."[415]

However, Saint Augustine of Hippo, an influential theologian of the early Catholic Church, argued that only God could suspend the normal laws of the universe. In his view, neither Satan nor witches had supernatural powers or were capable of effectively invoking magic of any sort. He wrote that it was the "error of the pagans" to believe in "some other divine power than the one God." Since witches were considered powerless, the early Church was not overly concerned with their spells or other attempts at mischief. The medieval Church accepted St. Augustine's view. For example, in the ninth century, Agobard, the Archbishop of Leon, dismissed the idea that witches could produce weather in a letter, "Against the foolish opinion of the masses about hail and thunder." [416]Early church documents went as far as to suggest that the belief in witchcraft itself was heresy.[417]

Looking back, European history from the Renaissance onwards is often written as a history of progress. The Renaissance, Reformation, and Scientific Revolution are often used to mark the stages of western society's emancipation from its medieval restraints. But when history is examined in depth, a more complex pattern emerges. Neither the Renaissance, the Reformation nor the Scientific Revolution was purely or necessarily progressive. It was during the Renaissance that Europe saw a revival of pagan mystery religion ideas and it would to take until the full maturity of the Scientific Revolution for these ideas to fade. And in between and beneath the surface of societies that were growing ever more sophisticated, dark passions were sometimes accidentally released and sometimes deliberately mobilized! The belief in witches is one such force. It was not a lingering ancient superstition only waiting to dissolve. It was a new explosive force that fearfully expanded with the passage of time.[418] The spark that led to the conflagration of the great European witch-hunts was ignited first when heretical groups like the Waldensians and the Cathars emerged as popular and competitive religious movements that threatened the power of the established church. The Cathar movement developed in the Langudoc area of what is now southern France and may have spread to parts of what is now Switzerland and Germany. Cathar dogma differed from the established church in that it described the human as having a spirit or spark of divine light within that was forced to live in an evil physical world that was also inhabited by the devil. The devil was portrayed as a powerful entity who was heavily involved in earthly affairs. In the Cathar world both God and Satan possessed supernatural powers. Cathar dogma described sexual intercourse as sinful. They

**Figure 34.** *Witches changing the Weather*, Ulrich Molitor,1493

## THE WITCH

avoided eating the meat of animals because they reproduced sexually and were the first sect to codify celibacy in their leaders. [419]

In 1208, Pope Innocent III opened an attack on the Cathar heretics that would serve as the opening salvo in a campaign that re-mystified the world of European Christian belief. The European public was given a crash course in the ways of the devil. Books, broadsides, paintings, rhetoric and torture were all part of the process. The devil, it appeared, had been set loose to walk the earth by these groups of heretics. The Church attempted to discredit the Cathar's beliefs by spreading stories about how they worshiped the evil deity in person. Propagandists for the church depicted Cathars as kissing the anus of Satan [the "kiss of the cat] in a ceremonial show of loyalty to him. It was a nasty encounter guaranteed to leave the participant with a spot or two on the skin.

When heretics responded to these threats from the established church by holding meetings in secret or after dark, a new stereotype of the witch's nighttime sabbat appeared for the first time. This blended known Jewish traditions and anti-Semitic sentiment into the mix. Some adherents of Catharism, fleeing the massacres of the Albigensian Crusades and the Inquisition that was launched against them, migrated into Germany and parts of territories that were newly under the rule of the Duke of Savoy.[420] When a Duke became suspiciously ill, Savoy became a geographic area where witches were heavily hunted. In fact, areas where Cathar influence was strongest were also most likely to be areas where more witches were "found."[421] Germany later also became the cradle of the Protestant Reformation.

This re-mystification campaign included various methods of torture which led to tawdry confessions. Defendants admitted to flying through the air to attend assemblies presided over by Satan, who appeared in the form of a cat or a goat-like animal. Some defendants admitted to casting spells on neighbors, having sex with animals, or causing storms. By the mid-thirteenth century it had became officially accepted that witches not only existed but were capable of causing physical harm to others and controlling the forces of nature. [422] The pagan folkloric concept of black magic had become fused with heresy to shape a distinctive crime of witchcraft. The public's understanding of Satan changed from that of a mischievous sprite to that of a deeply sinister force that could appear around any corner.

The first trials for witchcraft emerged out of the actions of the Inquisition, the official church-sponsored investigatory organization that was designed to stamp out heretical behavior. However, the history of

## DISGUISED AS THE DEVIL

systematic witch hunting is a more recent one. The significant period of witch hunting in the early (post plague) fifteenth century crystallized early modern witchcraft dogma. In 1485, Pope Innocent announced that in Germany, Satanists (possibly Cathars and Waldensian heretics) were meeting with demons, casting spells that destroyed crops, and aborting infants.[423] This outbreak of witch accusations coincided with a period of post plague reforestation and population growth in edge areas.

The Pope asked two friars, Heinrich Kramer and Jacob Sprenger, to create a full report on this suspected witchcraft. Two years later, the friars published *Malleus Maleficarum* or "Hammer of Witches" which put to rest, once and for all, the old orthodoxy that witches were powerless in the face of God. It promoted a new orthodoxy that Christians had an obligation to hunt down and kill witches. The *Malleus* told frightening tales about women who would have sex with any convenient demon, kill babies, and even steal penises. The *Malleus* became the principal text outlining the proper treatment of witches.[424] This book was instrumental in beginning to codify existing beliefs about witchcraft.

It also included for the first time a collection of a set of symptoms for demonic possession and witchcraft affliction. Witches were described with both specific and generalized characteristics that have prevailed in a somewhat modified manner to the present day.[425] Although eventually banned, it was an immediate best seller, over the next forty years the *Malleus* would be reprinted thirteen times. It was the bestseller of its day, second only to the Bible.[426] This new guidebook led the way to the stereotypical European witch and a wave of witch persecutions that lasted until the eighteenth century. It is during this period that the majority of the executions took place and the trials spread throughout Europe, Scandinavia and into New England.

Figure 35. *Malleus Maleficarum*. 1487.

# THE WITCH

The Reformation, which cleaved Europe into Roman Catholic and Protestant regions, only exacerbated the situation by causing social upheaval. A thoroughly re-mystified Europe retained witchcraft as a crime in both ideologies. Luther's Germany, full of the descendants of heretics and rife with sectarian strife, saw Europe's greatest witch execution rates. Witch hysteria swept France in 1571 after an accused witch announced that there were over 100,000 fellow witches roaming the countryside. Judges responding to the ensuing panic by eliminating most legal protections for accused witches. In his 1580 book, *On the Demon-Mania of Sorcerers*, Jean Bodin[427] opened the door to use of entrapment and instruments of torture on witches. Even then, there were some skeptics. Michel de Montaigne, who was invited to interview a French witch who he described as "an old woman, a real witch in foulness and deformity" whose malfeasance he attributed, however, "more to madness than malice...minds possessed rather than guilty."[428] Despite some skepticism, over the 160 years between 1500 and 1660, Europe saw between 50,000 and 80,000 suspected witches executed. Execution rates varied greatly by country, from a high of about 26,000 in Germany, to 10,000 in France, 1,000 in England, and only four in Ireland. Accused witches were disproportionately female.[429]

Early English folkloric witchcraft traditions included belief in the existence of cunning persons who could do good, or white magic, and those who committed black magic in the form of malicious malfeasance. These evil or black witches had to be guarded against. The work of witches came to be fairly well defined over time. Witches stole milk or caused it to go bad and could affect the outcome of the butter making process.[430] Witchcraft was associated with bad weather, especially floods and droughts that brought crop failure. In the Papal Bull that was placed as the introduction to the *Malleus*, Pope Innocent described the power of witches in Germany to

**Fig. 36.** *Kiss of the Anus, Compendium Maleficarum. 1608.*

## DISGUISED AS THE DEVIL

conjure droughts that destroy crops, writing:

"It has indeed lately come to our ears . . . many persons of both sexes . . . have blasted the produce of the earth, the grapes of the vine, the fruits of the trees, . . . vineyards, orchards, meadows, pasture-land, corn, wheat, and all other cereals . ."[431] Witchcraft was associated with infertility, disease, and affliction in both humans and domesticated animals.[432] As time went by the afflictions caused by witchcraft as well as descriptions of witches themselves were written down as part of historical records.

After a short lull, witchcraft trials in Europe re-emerged in the mid-sixteenth century with force. This may be related to the fact that the post plague population of Europe had grown large enough once again to intrude on the forest edge-restarting a deforestation cycle all over again. Indeed, when W. Behringer studied German witchcraft cases, he found that they were concentrated in poor and outlying agrarian regions. There is also an overlap with increased volcanic activity and the colder weather that ushered in the Little Ice Age. During this period, the numbers of trials generally rise and then fall as the ambient temperature falls and then rises. In addition to this general overlap, one of the sharpest drops in temperature in the Little Ice Age roughly coincides with a reinvigoration of witchcraft trials around 1560 after a 70-year lull.[433] When tree-ring evidence is compared with various non-dendroclimatic evidence, the body of evidence shows that a large region of mid and northern Europe experienced a sharp cooling at around 1570/80 that, at least in the north, marked a shift towards a prolonged period of cool conditions. This region had its southern boundary in the Alps. [There. There is little evidence for a major cooling in southern Europe.] In one Alpine area, this time period also corresponds with a drought.[434]

Drought years tend to concentrate ticks in riparian areas and along coastal plains.[435] This would be a time of vulnerability to Lyme disease in those areas and indeed the great witch-hunts in Europe overwhelmingly occurred in river valleys, near lakes, and in coastal areas. There is recent historical evidence that extreme weather was often a proximate cause of witchcraft accusations in Europe and North America.[436] A statistically significant relationship has also been found between extreme rainfall in the form of flood or drought and witch murders in modern Africa.[437] It is possible that these factors were the catalyst for the sixteenth century's resurgence of trials: the introduction and rapid spread of a new virulent *Bb* strain from the New World,[438] deforestation, drought years and cold

# THE WITCH

temperatures.

The history of the Massachusetts Bay Colony was also strongly influenced by these weather patterns. The famine and food shortages caused by the 1628 "year without a summer"[439] in England may have added special urgency to the Puritan need to immigrate to the New World. The Little Ice Age shaped New England's weather for years to come. But even this cold weather paled next to the prolonged drought that the colony suffered between 1680 and 1692. For twelve long years, rainfall in Massachusetts was well below average.[440]Such a drought was certainly within the sphere of the Anglo-witchcraft causal belief system.

**Figure 37.** *The familiar <u>Hunters in the Snow</u> by the Flemish painter Pieter Bruegel, the Elder, was completed in February of 1565 during the first of the many bitter winters of the Little Ice Age.*

The colony was surrounded by a forested wilderness full of indigenous groups that Puritans considered devil-worshipping heathens. The devil was always nearby in early New England. The use of fire as a weapon during warfare with these Native American groups sporadically added sunlight blocking smoke into the air that was already filled with light blocking dust by the volcanic activity of the 1680's that affected the atmosphere of the Northern hemisphere.

The *Compendium Maleficarum*, written in 1608, is a three book codified work "showing the iniquitous and execrable operations of

 DISGUISED AS THE DEVIL

witches against the human race." It includes a 47-item list for distinguishing those possessed by the devil and a 20-item list of the characteristics of the simply bewitched. The symptoms delineated fall into three general categories: experiential, fantastical, and religious dogma. The fantastical includes items like witches who go about the countryside collecting penises, flying on poles through the air and other such sort of testimony. The religious dogma includes belief about the power of holy water, the bible, and prayer to fight demons or cure.[441] It is the experiential lists, however, that bear particular interest for this study.

When the epistemology of witchcraft is examined, it is impossible to avoid the description of marks on the skin-both on witches themselves and sometimes on their supposed victims. Called alternatively witch's teats, witches marks, or devil's mark, they were thought to be brands made by contact or contract with the devil or made by the suckling of a demon that was usually in the form of an animal familiar. The Devil's 'making of a witch' caused a type of mark called a *stigmata diaboli* to appear. This red mark was also usually found in a hidden away place- under the arm, in the crease of the groin-in the very places often favored by ticks for attachment. It was also described as being insensate, much like the area that a tick numbs before and during attachment for a blood meal. These skin markings could take a variety of shapes. [442]

Witch marks are a sixteenth century phenomena that may have been created when a virulent form of *Bb* was introduced into Europe after the 1492 Spanish contact with North America. They are not mentioned in the *Malleus Maleficarum* of 1487, but become prominent features in witch literature after 1500. [443] Accused witches were strip-searched and unusual marks and protuberances were often found on their bodies. How were these marks made? The 1645 confession of the accused witch Margaret Bayts of Framington, England, holds a clue to the possible cause of the devil's marks that were found on her body. She confessed that "when she was at work she felt a thing come upon her leg and go into her secret parts, where her marks were found...and at another time when she was in the churchyard, she felt a thing nip her again in those parts." The devil's mark was described as being "in their secretist parts" where "he sucketh there." [444] The devil was thought to make the mark in a variety of ways- applying brimstone to the flesh, scratching with a cloven hoof or talon, but most often with a nip or bite of his teeth.[445]

A standard authority in legal procedure in England, recognized in witchcraft prosecutions in the New England colonies, was *Dalton's*

## THE WITCH

*Country Justice*, first published in 1619. The chapter on witchcraft contains a description of devil's and witch marks:

> "These witches have ordinarily a familiar, or spirit which appeareth to them, sometimes in one shape and sometimes in another; as in the shape of a man, woman, boy, dog, cat, foal, hare, rat, toad, etc. And to these their spirits, they give names, and they meet together to christen them (as they speak).... And besides their sucking the Devil leaveth other marks upon their body, sometimes like a blue or red spot, like a flea-biting, sometimes the flesh sunk in and hollow. And these Devil's marks be insensible, and being pricked will not bleed, and be often in their secretest parts, and therefore require diligent and careful search. These first two are main points to discover and convict those witches."

Witch marks and teats were searched for during the witch trials that were held in New England. Two descriptions of these teats exist from two cases in Connecticut. On Goodwife Clauson, "wee found on her secret parts[.] Just within ye lips of ye same growing within sid sumewhat as broad and reach without ye lips of ye same about on inch and half long lik in shape to a dogs eare which wee apprehend to be vnvsuall to women." In the case of Mercy Desboroughs, "and as to marcy wee find on marcy foresayd on her secret parts growing within ye lep of ye same a los pees of skin and when puld it is near an Inch long somewhat in form of ye fingar of a glove flatted..."[446]

However, those who were considered to be afflicted by witchcraft were also highly likely to develop red marks on their bodies. The testimony of the afflicted and the writings of eyewitnesses from New England are filled with descriptions of these marks- blisters, scalds, red streaks, and most importantly, bite marks of varying sizes. When the afflicted eight-year-old Elizabeth Kelley of Hartford, Connecticut was placed in her coffin in 1662, she had "a reddish tawny great spot, which covered a great part of the cheek."[447] In fact, many of the afflicted who are recorded in the Salem witchcraft literature from 1692 had some form of skin rash involvement. These were interpreted as being caused by contact with the invisible world-a burn from unseen brimstone, the result of contact with the devil in the disguise of a deer, or the bite of an invisible demon. A more scientific explanation for these marks is that they may have been the ECM's of Lyme infections.[448]

The cultural construction of a witch provides a description that is similar to the third stage of Lyme disease-the late or chronic stage. In general, a witch was an old woman living a marginal existence at best at

## DISGUISED AS THE DEVIL

the geographical edge of society itself. Prevailing cultural characteristics of witches were constructed as displaying a long list of symptoms. A witch was lame to the point of being stereotypically portrayed with a pole or crutch for walking. This pole would also be useful for flying through the air-certainly a less painful manner of travel for a crippled or disabled person. An accused witch [possibly Mary DeRich or Mary Hale] was described as "a woman who lives at Boston at the upper end of the town whose name is Mary. She goes in black clothes, has but one eye, with a crooked neck."[449]

Described as lean and deformed, witches were creatures of the night, wrinkled and pale from avoidance of the sunlight. A witch's vision was affected by a squint or "blear eye"- sometimes called the "evil eye," coupled with an inability to weep real tears. The eye of a witch, based on a prevailing *Doctrine of Effluvia,* could be used to inflict harm with invisible "venomous and malignant particles that were ejected from an evil eye into the bewitched to cast a fit."[450] Photophobia and optic neuritis accompanied by dry eye are sometimes symptoms of chronic Lyme disease.[451]

Another commonality in witch characteristics is an argumentative, antisocial and difficult personality. The personality of the witch was described extensively. "They are doting scolds, mad devilish, and not much differing from them that are thought to be possessed by spirits" showing melancholy in their faces to the "horror of all that see them." They are "miserable wretches" that are "so odious unto their neighbors and so feared as few dare offend them."[452]

**Figure 38**. *These witches are old, need canes and crutches to walk (or fly). The animals (all are blood meal hosts for ticks) depicted with them include a mouse, owl, cat, and dog (?) Woodcut, 1619.*

# THE WITCH

Martha Carrier was described by Cotton Mather as a "rampant hag." Suspected witches had a habit of muttering to themselves, a practice suggestive of the ability to curse others verbally and to communicate with the Devil and her familiars. The accused Salem witch Martha Corey, for example, would sit in front of her fire late at night talking to herself.[453] A condition called 'brain fog' often accompanies chronic Lyme disease-it interferes with both long and short term memory and causes confusion, poor word choice, and difficulty while speaking. The characteristics ascribed to a witch, such as the inability to recite the Lord's Prayer or the mixing up of words, may be the result of this *Borrelia* induced mental confusion.[454] Sometimes people accused of being witches must have sensed that there was indeed something wrong with them. In some cases, a neurological affliction may have caused them to confess. As Reginald Scott noted:

> "One sort of such as are said to be witches are women which be commonly old, lame, blear-eyed, pale, foul, and full of wrinkles; poor, sullen, superstitious...in whose drowsy minds the Devil hath gotten a fine fear so as, what mischief, mischance, calamity, or slaughter is brought to pass, they are easily persuaded the same is done by themselves..." [455]

Physician John Cotta was perplexed by the fact that sometimes the seemingly innocent voluntarily confessed to being witches. Even though they were "dying for sorcery," Cotta came to the conclusion that they also suffered from some form of illness that caused delusion and that they deceived themselves. Cotta states, "I grant the voluntary and uncompelled, or duly and truly evicted confession of a witch, to be sufficient condemnation of her self, and therefore justly has the law laid their blood upon their own heads, but their confession I cannot conceive sufficient eviction of the witchcraft itself..."[456]

Accused witches may have had another stereotypical characteristic. Although many appear to have been chosen arbitrarily, they may have

**Figure 39.** *Broadside of witches being burned, Germany, 1555.*

## DISGUISED AS THE DEVIL

actually been accused because they were or had been red haired. Although not specified in the bible, in medieval tradition Judas was thought to be a redhead.[457] Red hair is usually connected with freckled or spotted skin, which may have raised suspicions by their devil's mark like appearance.[458] In New England, Mary Glover, who was hung in Boston in 1688 as a witch for afflicting the Goodwin children, was a Gaelic speaking woman from Ireland, a country that has a significant population of redheads.[459] Others came from Atlantic fringe areas in Northern Europe that also have high levels of redheading. At least one accused witch, Rebecca Chamberlain, who died in prison in 1692 while accused of practicing witchcraft, may have been redheaded and has left redheaded descendants. In the areas of Europe where witches were burned, redheaded women may have made a particularly compelling case for being witches for another reason. Redheads have cells with a dysfunctional MC1R gene, which causes differential sensitivity to pain when compared with non-redheads. Redheads are especially sensitive to thermal pain, which may have produced a better "show" during the public events that witch- burning executions became.[460]

Folkloric cures and customs were part of the English oral tradition that passed from generation to generation through time and was imported into America. It is interesting to note that in English folkloric tradition some of the items that traditionally were used for protection against witches including bay leaves, ivy, ferns, and bracken, have insect repellent or insecticidal properties. The traditional ivy covered tower is mildly protected from insects and would be recognized in the seventeenth century as being well protected from witch-ly malfeasance.[461]

One of the treatments used for witchcraft affliction was included in Edward Drage's 1665 *Daimonomageia: A Small Treatise of Sicknesses and Diseases From Witchcraft and Supernatural Causes, Never before, at least in this comprised Order, and general Manner, was he like published. Being useful to others besides Physicians.* Writing about the case of Mary Hall, a "maid of woman's stature, a Smith's Daughter of Little Gadsen in the County of Hertford, [who] began to sicken in the fall of the Leaf, 1663. It took her first in one foot, with a trembling shaking and convulsive motion, afterwards it possessed both; she would sit stamping very much; she had sometimes like Epileptick, sometimes like Convulsive fits, and strange ejaculations…" Odd vocalizations occurred, as Mary began imitating cats, dogs and bears. It was thought that "spirits" spoke through her in an unnatural voice. She

## THE WITCH

was sent to "Doctor Woodhouse of Berkinsted, a Man famous in curing bewitched persons, for so she was esteemed to be; he seeing the Water [urine] and her, judged the like, and prepared stinking Suffumigations, over which she held her head, and sometimes did strain to vomit, and her distemper for some weeks seemed abated upon Doctor Woodhouse direction." The available medical practices for dealing with preternatural affliction included examining the urine, and applying smoke [possibly from burning feathers] to the head. [462]

There were numerous other folkloric methods for dealing with witchcraft that were practiced during the 1692 events in Salem Village. They provide a glimpse into the parallel folkloric belief system that, while being condemned by religious leaders, functioned in lockstep with both formal puritan religion and legal dogma among English colonists in America. Many involve fire, urine, and pins. At the suggestion of a Salem neighbor Mary Sibley, for example, the urine of an afflicted girl was used to mix up a "witch cake" that was then fed to a dog. It was believed that witchcraft affliction created invisible particles of effluvia in the victim that remained in the urine. When destroyed by the dog, supposedly, the witch would be forced to reveal herself. Another way to lure a witch was to burn or boil some part of an animal that had died from suspected witchcraft. This was thought to irresistibly attract the afflicting witch-the first person to come to your house afterwards was the guilty party. This is mentioned as being done several times during seventeenth century in New England.[463]

Protection from witchcraft was another matter. Cures and customs that were part of an oral folkloric tradition, passed from generation to generation through time, were in use in 1692. Mary Horado strung bay leaves on the door of her seventeenth century Berwick, Maine home to drive away witches.[464] Witch bottles were also probably in use. A relatively rare archaeological find in the New World, a few have been found in England, including the Riegate Witch Bottle. It was made in a style of glass that would have been familiar to the residents of Salem Village. The Riegate bottle was dug up at an archeological site just south of London, where it had been buried sometime after the date of the Salem Witchcraft Trials. The bottle was complete, the stopper was still in position, and it provided an unusual opportunity to examine the contents. The bottle's glass was made in about 1685, but it was not buried until some time around 1720. It had been well used before it was turned into a witch bottle and buried in a well-hidden spot to protect the maker's home. It was then discovered almost three hundred years later.

## DISGUISED AS THE DEVIL

When the contents of this bottle were analyzed, they were found to include bent pins, human hair (probably an eyelash) and urine. Pins in this case play an important role. Bending them may have represented protection by bending or breaking of one of the commonest symptoms of preternatural affliction-the neurological sensation of being pricked with pins and needles. The metal that the pins were made of may have also linked the maker to an even earlier English cultural belief set: the early Briton concept that iron and other metals had magical properties.[465]

When the first settlers left England for the New World they packed more than their material belongings onto their small wooden ships. They also brought along a full set of cultural concepts packaged tightly within their minds. These included a set of folk beliefs about sickness or affliction and witchcraft and religion. They were as capable of assembling a witch bottle as their counterparts back home or elsewhere in the sphere of Anglo influence. In fact, they may have felt as if they needed protection more than ever in a New World where the forests were inhabited by devil worshiping heathens in the form of Native Americans. [See Cotton Mather's statement in APPENDIX B] The few artifacts identified as witch bottles that have been found at sites in America show that both the idea and the practice had been transported across the Atlantic.[467] Witchcraft was a potent threat that generated great fear. On the sea, for example, it was believed that a witch could halt the wind and sink a ship. This folk belief led to the fact that several accused witches never even made it to dry land in the New World- they were hung at sea.[468] An analysis of existing records related to witchcraft in colonial America show some consistent themes that can be delineated as Anglo-folkloric beliefs.

**Figure 40**. *A 17th C. Bottle.*[466]

Witchcraft was comprehended to be a perversion of accepted religious doctrine. It was believed that the sacrament of baptism was perverted by the devil into a ceremonial immersion in sin. The records of the Salem witchcraft trials mention that Five Mile Pond and the Shawshin River were used by the devil for this practice, possibly because of their remote locations.[469] Social beings of a different sort, it was believed that a witch travels quickly,[470] often using chimneys as exits and flying on a pole through the air[471] to create mischief or to attend meetings with the

## THE WITCH

devil. [A modern persistence is that the entity known to fly through the air and use a chimney to enter and exit is the very magical but benevolent Santa Claus] Chimneys were necessary but dangerous as the entry point for malevolent creatures. [Think of the Big Bad Wolf of our nursery tales]People left empty shoes by fireplaces and hearths believing that they would catch evil spirits that came in through the chimney. [This practice has persisted as the benevolent placement of gifts in empty stockings and shoes that accompanies a visit from Old St. Nick] In 1692, witch meetings were testified as being held either on the open field near the Salem Village meeting house or at Andover in the Foster's pasture.[472] The evidence used to convict a person of being a witch included spectral evidence, confessions, and apparent proof of the witch's alleged supernatural abilities.

Fig. 41. *Witches exiting via a chimney.*[473]

The methods for obtaining confessions included pressing (Giles Corey), shackling hands to feet (the Proctor sons), beating (Roger Toothaker was probably beaten to death while imprisoned), the application of water to the face to simulate drowning, and trial by water. A witch was thought to float in water because the pure water would reject such an impure thing. This is again a throwback to Celtic belief in the magical powers of water that had been co-opted by Christianity. This trial by water was used in Connecticut in the case suspected witch, Mary Desboroughs, but the results were thrown out of court by a skeptical judge. [474] A witch was thought to have the power to leave their physical body or transform into a spectral apparition.[475]

The fertility of the world was perverted by witchcraft. Witches were ugly old hags with somewhat blurred sexual orientation. In the 1606 play, *Macbeth*, that was dedicated to the witch phobic King James I of England, the politically savvy William Shakespeare described a set of Scottish witches:

> "What are these so withered and wild in their attire....You should be women, and yet your beards forbid me to interpret that you are so."[476]

Witches took away the milk from cows.[477] Through spells, a witch was thought to affect the fertility of humans and animals-causing "offspring of women and the foals of animals to perish" or behave

## DISGUISED AS THE DEVIL

strangely. Crops could be mysteriously destroyed.[478] A witch was thought to afflict and torture with preternatural fits, pinching, pricking (with unseen pins), and cause dire pains and anguish, both internal and external, to men, women, children, cattle, flocks, herds, and animals.[479] Horses could be made to behave unnaturally.[480] Witchcraft attacked the senses, caused blindness, deafness, and loss of speech. Children were made lame. Children were made hyperactive. Witches made humans bark like dogs or cluck like chickens. Some of the afflicted even thought that they were flying.[481] In one case in Salem, the witch (Susannah Martin) was described as becoming a will-o-wisp to lead a man astray in the woods.[482] In the most extreme circumstances, the witch could cause death.[483] Witches were creatures of darkness, inhabitants of the night. Daylight was anathema to a witch. It caused them to squint-the classic mark of the evil eye.

The affliction that witchcraft caused in humans was, as Cotton Mather remarked, outside the realm of the normal. It is important to understand the mental concepts applied by the seventeenth century doctors who played an important role in the identification of witchcraft-associated cases in New England. The role of the physician's diagnosis of witchcraft was significant. Physicians and surgeons were the principal professional arbiters for determining natural versus preternatural illness. Based on medical writing of the seventeenth century, New England's physicians were trained to believe in the possibility of preternatural causes for disease. Medical writers of the time like John Cotta, Edward Jorden, and William Drage delineated the nature of certain rare, infrequently seen, odd and puzzling symptoms in patients that could possibly be caused by witchcraft. Many of the witnesses or afflicted involved in witchcraft cases noted in their depositions that they saw a particular doctor or several doctors who had difficulty diagnosing their baffling illnesses.[484] This is number one on the list of *The Compendium Maleficarum's* commonest "signs which show a man to be simply bewitched." It reads: "the patient's sickness is very difficult to diagnose, so that the physicians hesitate and are in doubt and keep changing their minds, and are afraid to make any definite pronouncement about it."[485]

The suspicious illnesses linked to witchcraft had several aspects in common. These included the suddenness of the onset of the symptoms, the strangeness of the symptoms, and the rarity of the complaint. John Cotta argued that there were few ways to distinguish natural from diabolically caused disease in the afflicted. The first was the presence of

## THE WITCH

an extraordinary, wholly unexpected or contrary symptom that did not fit in with other symptoms. The second way was to note the patient's reaction to medications that had a predictable physiological effect on the body. If the medications did not have the usual results or produced wholly unexpected effects, witchcraft was a likely diagnosis.[486]

The seventeenth century practitioner also had to take what was called dissembling (lying) into account. They knew that some patients faked or exaggerated their symptoms. The charge of dissembling was certainly leveled against the afflicted of Salem Village and there may have been some truth to the charges. Mary Warren reputedly told Elizabeth Hubbard that "the afflicted persons did dissemble." Daniel Elliot testified that he overheard one of the afflicted say that "she did it for sport. They must have some sport."[487]

Despite such testimony, the vivid impression of the fitful agony of the afflicted in 1692 and the comparison of symptoms to earlier cases convinced contemporary observers that the illnesses were very real. Nathaniel and Hannah Ingersoll testified that right before his death Benjamin Holton was behaving "in a very strange manner with most violent fits acting much like to our poor bewitched persons."[488] Distraction [mental illness] was another possible diagnosis. This was defined as a state of disordered reasoning, delusional behavior, and melancholy. New England's early doctors diagnosed individuals whose mental functioning was chronically impaired and whose thoughts were scattered, illogical and unfocused on a continuing basis as distracted. Most of the afflicted in the Salem Trials did not fit this pattern. The afflicted girls were often lucid, and their accounts of their experiences of spectral visions or hallucinations were well within the period's cultural boundaries of what was considered believable.[489] Hysteria was thought to be caused by the movement of the womb or "the mother" through the abdominal cavity. The problem that the seventeenth century physician had with using this diagnosis in Salem was that hysteria was not considered to be an infectious disease. The fact that several individuals manifested the same or similar hysterical symptoms at the same time made it a 'remarkable' occurrence[like the group of arthritic children in modern Lyme].[490]

But many physicians must have felt as puzzled as Cotton Mather, who wrote 'I think I may, without vanity, pretend to have read not a few of the best systems of physick that have yet to be seen in these American regions, but I confess that I have never yet learned the name of the natural distemper, whereto these odd symptoms do belong."[491] The answer to his

 DISGUISED AS THE DEVIL

puzzlement may lay outside the bodies of the afflicted-in the landscape of seventeenth century Salem Village itself.

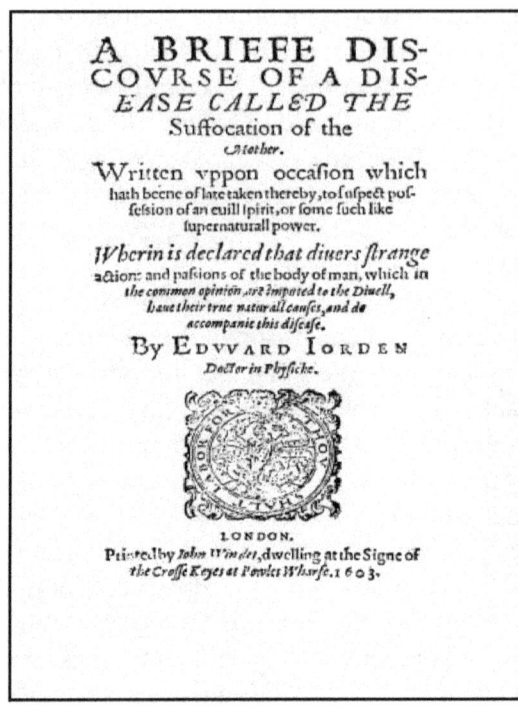

**Figure 42.** *Frontispiece, Edward Jorden, A Disease called the suffocation of the mother, 1603.*

# THE WITCH

## THE WITCH AT THE EDGE OF THE WOODS

410 See King *James Bible*, Exodus 22.18, Leviticus 20:27, Deuteronomy 18:10-11.
411 From Reginald Scot's *The Discovery of Witches* (Mineola, N. Y.: Dover Publishing, 1989), originally printed in1595, which was openly skeptical of the existence of witchcraft. All obtainable copies were burned on the accession of James I to the English throne in 1603 and those remaining are now extremely rare.
412 Michael Bailey. "The Disenchantment of Magic: Spells, Charms, and Superstition in Early European Witchcraft Literature," *American Historical Review* Vol. 3 No. 2, April 2006.
413 Josiah C. Russell, "Population in Europe" in Carlo M. Cipolla, Ed., *The Fontana Economic History of Europe, Vol. I: The Middle Ages* (Glasgow: Collins Fontana, 1972), 25-71.
414 Thomas van Hoof, Frans Bunnik, et. al. "Forest re-growth on medieval farmland after the Black Death: Pandemic-Implications for atmospheric $CO_2$ levels," *Palaeogeography*, Palaeoclimatology, Palaeoecology 237 (2006) 396-411,uses well dated pollen assemblages to reconstruct regional vegetation and land use for the period between 1000 and 1500 AD in the Southeastern Netherlands.
415 King James *Bible*.
416 Agobard of Leon, *Contra insulsum vulgi opinionem de grandine et tonitruis* was written in the 9th century A.D.
417 Alan Charles Kors, and Edward Peters, ed. *Witchcraft in Europe 400-1700: A Documentary History* (Philadelphia: University of Pennsylvania Press, 2000).
418 Hugh Trevor-Roper, *The Crisis of the Seventeenth Century: Religion, the Reformation and Social Change* (Indianapolis: Liberty Fund, 2001). Accessed from http://oll.libertyfund.org/title/719/77029 on 2008-03-19.
419 Jeffrey Burton Russell, *Witchcraft in the Middle Ages*, (Ithaca: Cornell University Press, 1972), 17.
420 Joseph R. Strayer, *The Albigensian Crusades* (Ann Arbor: University of Michigan Press, 1992).
421 Russell, Witchcraft, 124.
422 See Francesco Maria Guazzo, Milan, *Compendium Malfaricam* (Mineola, N.Y.: Dover Publishing, 1988) was originally printed in Milan in 1608.
423 Wolfgang Behringer, *Witches and Witch Hunts* (Cambridge UK: Polity Press, 2004), 72.
424 Summers, Montegue, Ed., *The Malleus Maleficarum of Heinrich Kramer and James Sprenger* (Mineola, N.Y.: Dover, 1971).
425 The devil's mark is not yet a characteristic of a witch in the Malleus in 1487, by the Compendium Malfaricam of 1608, it is a characteristic trait, which may indicate the introduction of a virulent ECM producing strain of *Bb* from the New World by Spanish explorers.
426 Behringer, Witches, 71-74.
427 Jean Bodin, *On the Demon-Mania of Sorcerers* (U.K.: Centre for Reformation and Renaissance Studies, 1995).
428 William Hazlitt, Ed., *The Essays of Michel de Montaigne*, Book the Third, Chapter XI Of Cripples, 1877, Project Gutenberg e-book #3600.
429 See Behringer, Witches, Carol Karlsen, *The Devil in the Shape of a Woman: Witchcraft in Colonial New England* (New York: W.W. Norton, 1987) also discusses this phenomenon.
430 Butter making is greatly influenced by the amount of humidity in the air. From my experience as a historical interpreter, I can confirm that trying to churn butter on a humid day is indeed a frustrating experience. See also Henry Glassie's description of Irish witch beliefs in *Passing the Time in Ballymalone*.
431 Malleus (1487) Papal Bull as Introduction.
432 Ernest Baughman, *Type and Motif Index of Folktales of England and North America* (1966) motif 249
433 See Behringer, W. "Climatic Change and Witch hunting: The Impact of the Little Ice Age on Mentalities," *Climate Change*, 43(1999), 335-351, see also Christian Pfister, Rudolf Brázdil. "Climatic Variability in Sixteenth-Century Europe and its Social Dimension: A Synthesis," *Climatic Change*, Volume 43, Number 1 (September 1999), 5-53,
434 Historic climates have been recreated using tree rings to predict the Palmer Drought Severity Index- Europe. See: K.R. Briffa, P.D. Jones, R.B. Vogel, F.H. Schweingruber, M.G.L. Baillie, S.G. Shiyatov, E.A. Vaganov, "European Tree Rings and Climate in the 16th Century," *Climatic Change* Volume 43, Number 1 (September 1999), 151-168.
435 See Jones and Kitron study, "Population .... modulated by drought."
436 Behringer, *Witches*, 344 and Oster, E., "Witchcraft, weather and economic growth in Medieval Europe," *Journal of Economic Perspectives* 18 (1) (2004), 215-228.
437 Miguel, E, "Poverty and witch killing," *Review of Economic Studies* (2003), 1171.
438 Wei-Gang Qiu, J. Bruno, W. McCaig, Yun Xu, Ian Livey, M.Schriefer, and B. Luft, "Wide distribution of a High-Virulence Borrelia burgdorferi Clone in Europe and North America," *Emerging Infectious Diseases*,

Vol. 14, No. 7, July 2008, 1097-1104.
439 Behringer, Witches, .xv. 1628-'year without summer.'
440 See NOAA: World Data Center for Palaeoclimatology at www.ncdc.noaa.gov/paleo/treering.html
441 Guazzo, *Compendium Maleficarum(Milan,1608)*.
442 Ibid.
443 Richard M. Golden, *Encyclopedia of Witchcraft: The Western Tradition*, Vol. 4, Q-Z, (Santa Barbara, California,:ABC-CLIO, 2006).
444 Rossell Hope Robbins.,*The Encyclopedia of Witchcraft and Demonology* (New York: Crown, 1963), 552.
445 Rosemary Ellen Guiley, *The Encyclopedia of Witches and Witchcraft* (New York: Facts On File, 1989), 99.
446 John Taylor, *The Witchcraft Delusion in Colonial Connecticut 1647-1697*(1908) electronic edition available online at www.fullbooks.org.
447 u.k., "The Earliest Recorded Autopsy in America Performed in 1662 on the 8-year- old Elizabeth Kelley," *Pediatrics* Vol. 61 No. 4 (April 1978), 572.
448 While there is debate about the level of ECM appearance in Lyme disease because the only sample in most studies is a sample of convenience composed of people who come in to a physician to be treated often because they find an ECM. People without ECM's may be less likely to seek treatment which skews the findings of most of the available studies. These studies estimate that between 30 and 80 percent of all cases of Lyme disease in the United States show this rash which may overstate its prevalence. However, while it does not always appear, it does appear some of the time.
449 There were a lot of accused witches named Mary. This might be Mary De Rich, who was accused of witchcraft for afflicting Elizabeth Booth, Abigail Williams and Elizabeth Hubbard in 1692. Essex County Archives, Salem Witchcraft Vol. 2. 51, but her residence is listed as being in Salem Village not Boston. Mary Hale of Boston is another possibility.
450 William George Black, *Folk Medicine: A Chapter in the History of Culture* (London: The Folklore Society, 1883), 21-23.
451 See Mikkila, H.O., et al. "The expanding clinical spectrum of ocular Lyme disease," *Ophthalmology*107, 3(2000), 581-7.
452 See Robin Briggs, *Witches and Neighbors: The Social and Cultural Context of European Witchcraft* (New York: Viking, 1996).
453 Giles Corey testimony, Essex County Archives, *Salem Witchcraft*, Vol. 1, 13.
454 George Burroughs passed the Lord's Prayer test with flying colors but was executed anyway in 1692. A witch complaint from Virginia in 1703 states that the accused witch mixed up words while speaking as a sign of being a witch. There are times when I cannot remember my pin number in the grocery store-I would have been doomed in 1692. I get annoyed when news commentators focus on the fact that President Bush seems to sometimes have difficulty finding the proper word or gets a bit mixed up with the order of his words while speaking, which makes for amusing sound clips but may in fact be related to the lingering effects of his bout with Lyme disease.
455 Reginald Scot, *The Discovery of Witches*.
456 John Cotta, *The Triall of witch-craft Shewing the True Methode of the Discovery*, (New York: Da Capo Press, 1968) Reprint.
457 The fact that Judas was the subject of one of Rev. Parris's sermons made during the Salem Witchcraft outbreak is perhaps more than a coincidence.
458 Marion Roach, *The Roots of Desire* (New York: Bloomsbury USA, 2005).
459 She needed an interpreter at her trial for witchcraft because she spoke Gaelic. See Norton, In the Devil's Snare, 39 and Hansen, Witchcraft, 23.
460 E.B. Liem, Joiner TV, Tsueda K, Sessler DI. "Increased sensitivity to thermal pain and reduced subcutaneous lidocaine efficacy in redheads," *Anesthesiology* 102(3) (March 2005), 509-14.
461 Fearnley-Whittingstall, Jane. *Ivies* (London: Chatto & Windus Limited, 1992).
462 Stinking suffumigations were sulperous smoke created by burning feathers or hair.
463 See Burr, *Narratives*, 239. Modern scientists have begun to take a second look at how diseases affect urine. See Nigel Hawkes, "Scientists discover drops of truth in medieval belief in urine," *The Times of London*, April 21, 2008. See also Elaine Holmes, Ruey Leng Loo, Jeremiah Stamler, et al, "Human metabolic phenotype diversity and its association with diet and blood pressure," *Nature*, published online April 20, 2008.
464 Emerson Baker includes an interpretation of the H(F)otado witchcraft incidents as a case of severe domestic abuse in *The Devil on Great Island*, a wonderful book about pre-1692 witchcraft accusations in Northern New England, an area that, like Virginia, is generally neglected in witchcraft literature. The use of Evergreen (bay) as protection from witches is mentioned on page 166. Our Christmas decorations of evergreen, boughs of holly and mistletoe may be derivative of these folkloric traditions.
465 Article about Alan Massey, "How to Kill a witch: The Reigate bottle" in *Current Archeology*, August

# THE WITCH

2000 at www.archaeology.co.uk/ca/issues/ca169/witch/witch.htm
[466] Collection of the author
467 See Marshall Becker, "An Update on Colonial Witch Bottles," *The Pennsylvania Archaeologist* Vol.75 Issue 2, Fall, 2005, and also "American witch bottle," Archaeology Vol. 33 No.2 (March/April 1980).
468 Mary Lee and Katherine Grady were both hung at sea in1654 before they reached Virginia. See Richard Beale Davis, "The Devil in Virginia in the Seventeenth Century," *Virginia Magazine of History and Biography* 65 (1957), 131-49.
469 See Testimony of William Barker, *Suffolk Court Records* Case No. 2761, 102 and George Lincoln Burr, ed., *Narratives of the New England Witchcraft Cases* (Mineola, New York: Dover publications, 2002 Reprint), 189,241.
470 Ernest Baughman, *Type and Motif Index of Folktales of England and North America* (1966) motif 249.
471 Robert St. George, *Conversing By Signs*. The chimney is now used by a magical Santa Claus to enter and exit homes on Christmas Eve.
472 See Testimony of William Barker, *Suffolk Court Records* Case No. 2761,102 for a description of this activity.
[473] Gillotole de Givry, 1579.
474 For the standards used to examine witches in Connecticut see Appendix F, and Burr, *Narratives*, 21n.
475 Spectral evidence appears frequently through out the Witchcraft trial testimony, see especially, Ann Putnam and Elizabeth Parris testimony in P. Boyer and Stephen Nissenbaum , *The Salem Witchcraft Papers*, Vol 2.
476 William Shakespeare, *Macbeth*, Act I.III lines 39-46.
477 See testimony of John Roger (Essex County Archives, Salem Witchcraft Vol.1), 139.
478 Papal Bull in *Malleus maleficarum*, 1484, sets the precedent for these witchcraft accusations in Europe.
479 This is another theme that runs throughout the Salem testimony.
480 Witchcraft motif G 265.6.3 in Thompson, *Motif Index of Folk Literature* and Isaac Cummings Testimony, Essex County Archives, *Salem Witchcraft* Vol.1, 148.
481 The description of these symptoms throughout testimony see also Cotton Mather *Memorable Providences*, 1689, and *A brand Pluck'd Out of the Bur*ning, 1693, in Burr, *Narratives*.
482 Cotton Mather , "The Wonders of the Invisible World" (1693) in Burr, *Narratives*, 235.
483 See testimony of The Rev. John Hale and Sarah Holton testimony in P. Boyer and Stephen Nissenbaum, *The Salem Witchcraft Papers*, Vol 2.
484 Lawrence B. Goodhart, "The distinction between witchcraft and madness in colonial Connecticut", *History of Psychiatry*, 13 (2002), 433-444.
485 Compendium list.
486 John Cotta, *The Trial of Witchcraft Showing the True and Right Method for the discovery with A Confusion of Erroneous Wayes*, (London: George Purslowe, 1616).
487 Goodhart, The distinction.
488 See testimony of The Rev. John Hale and Sarah Holton testimony in P. Boyer and Stephen Nissenbaum, *The Salem Witchcraft Papers*, Vol 2.
489 Goodhart, *The distinction*.
490 Edward Jorden, *a Disease called the suffocation of the mother* (London: John Windes, 1603).
491 Cotton Mather, *A Brand* in Burr, *Narratives*.

# DISGUISED AS THE DEVIL

SALEM VILLAGE

# VIII. SALEM VILLAGE, MASSACHUSETTS:

In seventeenth-century Massachusetts the landscape gradually developed an overlay of the physical, social, and cultural mindset of the new English settlers, first filling in land that had previously been cleared by a deceased Native population. Any attempt to give population figures for these colonies during the seventeenth century is an estimate. The number of settlers who immigrated to the Plymouth Colony is estimated to be a total of three hundred and sixty two.[492] The estimates for migration into the Massachusetts Bay Colony between 1628 and 1640 (when it came to an abrupt end because of legal mandates in England) vary widely from 10,000 to 20,000. The colonies in New Hampshire and Connecticut were also largely populated from this "great migration."[493]

The estimates of the pre-disease Native American population for this same land area also vary widely. Daniel Gookin, writing during the seventeenth century, sets the Massachusetts tribe's pre-contact strength at 10,000 but there are questions about which groups he included or excluded in that number. Others have given estimates of 20,000 or more.[494] Although Native and English land use patterns would have differed, it appears that a roughly equal number of English settlers may have replaced a similar number of deceased Natives on the same land. Since Native activity and land use often included using land until it was "worn out" and then moving on by clearing new fields for cultivation, it is quite likely that the English found fairly extensive amounts of cleared land to settle before they took to the forest in force.

It is highly likely that after an initial period of exploration, clearance of accumulated brush and undergrowth on the sites chosen for habitation, and building of homes and fences, the first colonists chose to interact infrequently with the forests. They preferred to settle on land that was already cleared of trees, and avoided the dangers of wolves, hostile Native Americans, or getting lost that the wilderness represented. The business of the early seventeenth-century settler lay in cultivated places on the very fringe of the forest, sometimes selectively extracting pine and oak trees to use as timber for houses and for shipbuilding. This created a dispersed group of settlements in New England that tended to be along rivers with dense forest in between. Waterways were the great highways of seventeenth-century Massachusetts. Looking for good cleared land, the first group of settlers that went from Massachusetts to Connecticut went by sea. Travel between Plymouth and Boston was by water. Travel between Boston and Salem was often by water. One early observer commented on the frequent use of canoes for transportation in the Salem area, writing, "there be more canoes in this town than in the whole patent, every household having a water horse or two." [495]The typical early seventeenth century New England town either fronted the ocean or faced in on itself with a river nearby.

    Because the English came from a crowded and increasingly land cramped island and practiced a system of manuring or "dressing" fields to promote continual fertility, along with a rotating set of pastures for horses and cows, even by the end of the "great migration" the English had probably not even begun to occupy all the cleared spaces on the map or intrude extensively into the forest yet. But by the 1650's and 60's the population on the outer edges of the Massachusetts Bay Colony began to have contact with the wilderness; forced by the need for more lumber, fields for crops and pasturage, clearing and fragmenting the forest in the Salem Village area, and intruding into the forest that ringed the frontier settlements of Andover and Billerica. The shipbuilding industry needed oak for timber and pine for masts. Fledgling iron and copper works burned prodigious amounts of charcoal to process local metals.[496] Potash and firewood were needed on a daily basis for cooking, soap making, and tanning leather.[497]

    Historic Salem Village was located on the land that is now the modern town of Danvers, outside the town of Salem. Like Salem itself, the village was built partially on land that had been cleared by Native Americans. This area, called Naumkeag, was described by John Smith as having "a multitude of people." He wrote that the "sea coast as you passe, shewes you all along large corne fields, and great troupes of well proportioned people."[498] By 1628 the multitude of well proportioned Native Americans were dead, all but eliminated in waves of plague and pestilence that had occurred before the first English settlers even arrived. At contact, the peninsula that is now Salem, the Northfields area, and a few outlying spots in Salem Village were clear of trees.[499] The first English settlers in the area were a combination of the two hundred member Higginson Fleet of emigrants mostly from the East Anglia area that arrived in 1628 and about fifty men of west country origins (Dorset), including John Endicott and Roger Conant, some moved down from a fishing outpost that had been located earlier on Cape Ann.[500] Francis Higginson described the landscape at contact. "Though all the countrey

## SALEM VILLAGE

be as it were a thicke wood for the general, yet in divers places there is much ground cleared by the Indians and especially about the plantation." The soil was "in other places clay... in other gravel, in other sandy [as it is about our plantation at Salem]."[501] The thick wood that Higginson noted, as is shown from pollen core sampling, was dominated by oak and pine trees.

The site of Salem was chosen because it had access to the water, a good harbor, and was defensible against attack. A fort was built to protect the entrance to the town on the land side, and during King Phillip's War a palisade was added, stretching across the base of the peninsula as protection from Native American attacks. While the land on the peninsula was not especially fertile, it was used for small gardens in the immediate vicinity of the first English houses. The mainland adjacent to the Salem peninsula featured salt and fresh marshes that were used almost immediately for the common grazing of cattle. This common land was called the Northfields. Beyond this zone, outside town, in what was to become the east side of Salem Village, there was an especially fertile plain. A few Native paths traversed the area and would become over time the roads that were used by the English settlers to travel between Ipswich and points to the north and Boston to the south. [502]

Several land grants for the area outside town were made beginning in the 1630's. The first, in 1632, was for three hundred acres to Governor John Endicott for what became Orchard Farm. Another two hundred acres were later added to the grant in exchange for improvements to be made to the Ipswich Road. Land in the adjacent area was parceled out to several different settlers.[503] In 1639 the General Court of Massachusetts Bay Colony defined the outer limits of the village by stating that "wheras the inhabitants of Salem have agreed to plant a village near the river which runs to Ipswich, it is ordered that all the lands near their bounds, between Salem and the said river, not belonging to any other town or person by any former grant, shall belong to said village."[504] Immediately afterwards, grants were given for land on the Ipswich River, described as "a hill, an Indian plantation, and a pond" on the western side of the forest that included "about one hundred or one hundred and fifty acres of meadow." Thomas and Nathanial Putnam bought one hundred and forty acres of upland and fourteen acres of meadow in this western zone. Bray Wilkins also bought land in this area in 1661.[505]
The village grew from these two areas-in from the east near Orchard Farm and then along the Ipswich Road, and also inward from the western Ipswich River edge towards the east. The wooded center of the village

 DISGUISED AS THE DEVIL

only began to fill in during the 1660's and 1670's in a splotchy crazy quilt development pattern that was created by the English land ownership system. As time went by, large land grants were divided up between the descendents of the original owners or pieces were sold. Farms were interspersed with woodlots, meadows, and various industries. In 1667, the village was described as a group of "houses distant from each other, some ten miles, some eight or nine miles from the town of Salem." The distance of the houses, one from another were "some a mile some further- that it is difficult sending a neighbor to another in dark nights in a wilderness that is so little cleared and [by] ways so unpassable." This wooded wilderness was a dangerous place, especially "considering what dreadful examples former times hath afforded in that respect, in this country, from Indians... in the night season." [506]

Service occupations like a doctor, a plate turner, the owners of public houses and the grist miller clustered near the eastern zone, especially along the Ipswich Road. Other, more forest consuming activities occupied residents of the Salem village and Andover areas. The Putnam family started an iron processing operation that was located north of the village. Governor Endicott owned a copper mine and processing facility somewhere on his land that processed eight tons of ore per week.[507] Several sawmills were established along streams in the western zone to process wood. [508]

Deforestation began slowly, often with the selective extraction of oak. Houses, furniture, ships, barrels and fence rails were all made from this hardwood. The Endicott Orchard Farm, for example, was enclosed by a palisade that took 7,000 six-inch wide trees to build.[509] The iron operation set up north of town by the Putnam family would have burned five cords of wood (converted into charcoal) for each ton of iron that was produced. Another tree consuming set of iron works was set up in north of the village in Andover, on the Shawshin River, by the Chandler family.[510] An acre of trees had to be cut for every two tons of potash that was produced and used as fertilizer or to create lye. Tanners, like the Parkers of Andover, and dyers, like William Shattuck of Essex Street in Salem Town, used bark for hide processing and coloring cloth and wool. The forest's tall and straight pine trees were cut for use as ship masts. Others were bled for their resin to make pitch, turpentine and other naval wares.[511] Cordwood went to Salem's brick kiln, limekiln and its biscuit and bread ovens. Once homes had been built, they required prodigious amounts of wood for cooking and heating. When the Salem Village church hired its

# SALEM VILLAGE

first minister, it contracted to supply him a total of thirty cords of wood for use in the new 20 by 30 ft. parish house.[512]

This deforestation, when combined with an English pattern of fixed land ownership, created a mosaic of farm, ecotone edge areas and forest fragments that can be very clearly be seen on the map. From the unwanted swamp Blind Hole in the southwest, a patchwork quilt of ownership and land uses spreads east across the land showing lots of edge environments, orchards, and riparian zones.[513] When compared to modern maps, soil surveys and Geographical Information System vegetation surveys, many aspects of this landscape were particularly likely to harbor ticks and a high risk for potential infection with Lyme disease.

The Salem Village growth pattern can be seen in some of the relict features of the Salem Village area that remain in modern Danvers. Some remaining structures represent successional growth and expansion patterns. These include the Joseph Putnam House, built in 1648, located in what was part of the west to east growth zone. The Endicott burial ground (which was a Native American camp before it became a cemetery) is in the east to west growth zone and has early graves that are from the 1650's. The Daniel Rea House (1660), the Samuel Houlton House (1670), the Joseph Holten House (1670), and the Benjamin and Sarah Houlton House (1670) are all part of a wave of growth in the center of the village. By 1671, Watch House Hill had been fortified and a militia training field was designated nearby. In 1672, Salem Village became a separate parish and a meetinghouse was built. The center of the village became more heavily occupied. A third wave of growth is represented by the house that was occupied by Rebecca and Frances Nurse, built in 1678, the Osborn House of 1680, the Haines House of 1681, the Zerubabel Endecott House of 1681, the (Parris) Parsonage of 1681, the Putnam Perry House of 1685, and the John Holten House of 1692.[514]

The first settlers on the great meadow at Andover, just to the north of Salem Village, and in Billerica to the west, also occupied a mosaiced landscape along the Shawshin and Ipswich River basins. In addition to grassland, the area contained "woods" interspersed with grassy clearings, several hills, numerous creeks and rivers, meadows, swampy areas, rocky soil and several ponds that had been left behind by the retreat of the last Ice Age glacier. This was all land that "pleased not" the Winthrop fleet when they first arrived and sent them to Boston in search of better land. It also seems to have pleased not some of the first settlers. John Winthrop Jr. lived in Salem Village for a very short time before moving on to

Saybrook, Connecticut (a move predicated in some areas by the sickness of cattle).[515]

A survey report to the Massachusetts General Court described the Shawshin area (the Shawshin River flows through Billerica towards Andover) as land that is "no way fit, the upland being very barren & very little meadow there about, nor any timber almost fit for use... between the side of Concord line & the head of Cambridge line, but little meadow & the upland of little worth." The most prominent features seem to be swamps, poor land, and a spider web of streams, brooks and riverlets.[516] To settle one of many property disputes between neighbors in Salem Village, surveyors had to trek through a swamp and cross three brooks to measure the land.[517] Even in 1692, a large section of the western part of Salem Village was dominated by a swamp and appears to be unused. Andover had Wolf Swamp, located next to Deer Meadow, and everything was edged by the forest to the north.[518]

Salem may have been unique in that it was settled by two groups – a party of old planters who came from West Country (the Dorset area) before 1628 and a contingent of East Anglians who arrived in the great migration during the 1630's. They brought with them two very different ideas of community that differed in material ways. Salem's West Countrymen had originated in an area of England that had dispersed and separate farms. East Anglians saw the ideal community as the compact villages they had occupied in England. After 1636, the West Countrymen left the town on the peninsula and spread out into the village area, building farms and hamlets in a dispersed pattern that was familiar to them. The East Anglians, for the most part tended to stay in town.[519]

As the population began to spread westward and eastward they moved into areas that had brushy, weedy edge zones along the Ipswich River and forest edges--the wet green zones that are risky for Lyme disease. They then proceeded to fragment the forest into woodlots and open areas in a dispersed pattern- these became improved habitats for mice and ticks, especially when there were pig population fluctuations, and sporadically people and animals became very sick.

**MAP 9: Salem Area Growth and Land**

# SALEM VILLAGE
## Development: 1628-1692

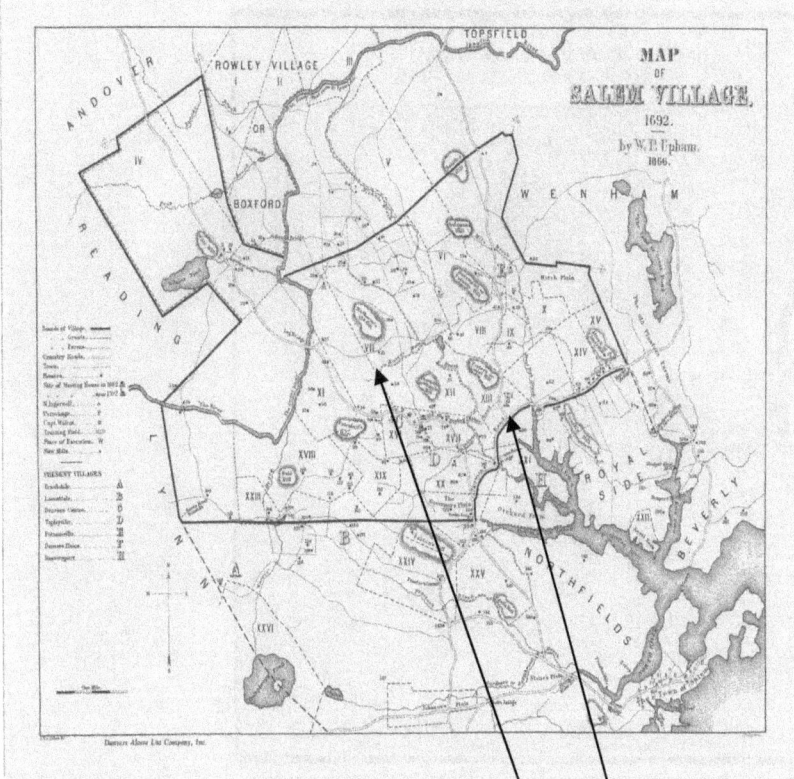

**Early** Settlement-1639-40's in west, 1630's in East
**Second Wave** of Growth is clustered up the Ipswich Road, 1640's in East, timbering and Iron Operations in the West, 1650's
**Third Wave** of Growth fills in the center in a dispersed pattern, the meetinghouse is built, training field created for militia, and watch tower built, 1660-1692.
**Riparian Zones**: Blind Hole Swamp, The Ipswich River, various streams and creeks crisscross the area. **Woods**: Lots without houses, sides of stony hills, south of Village Line, to the west and north.

# Putnam Farm Landscape and Soil Area:

 DISGUISED AS THE DEVIL

# Pre-1628

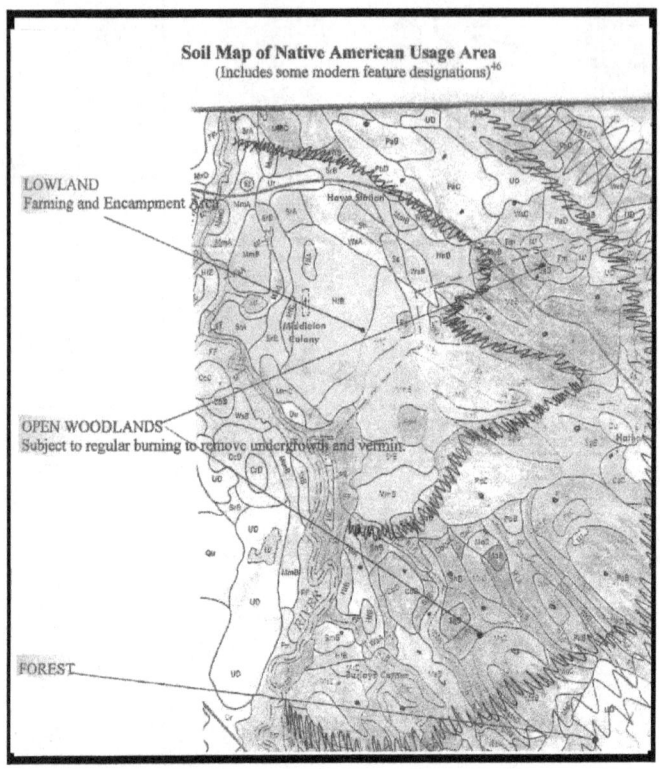

**Map 10:** [520] shows how the Native American settlement in the area had been situated. Native Americans tended to live and farm in rich lowland areas. This occupation zone was surrounded by an upland open forested buffer zone that was subjected to regular burning that cleared both undergrowth and vermin. This zone was in turn surrounded by heavily forested areas. This configuration would have a **LOW RISK** for Lyme disease. The Native Americans who had lived at this camp died in the "great sickness" that preceded English settlement in New England. This same area was settled by members of the Putnam family starting in the 1640's and over the course of time as more forest was cleared was split into several family farms.

## Putnam Farm Area: 1692

# SALEM VILLAGE

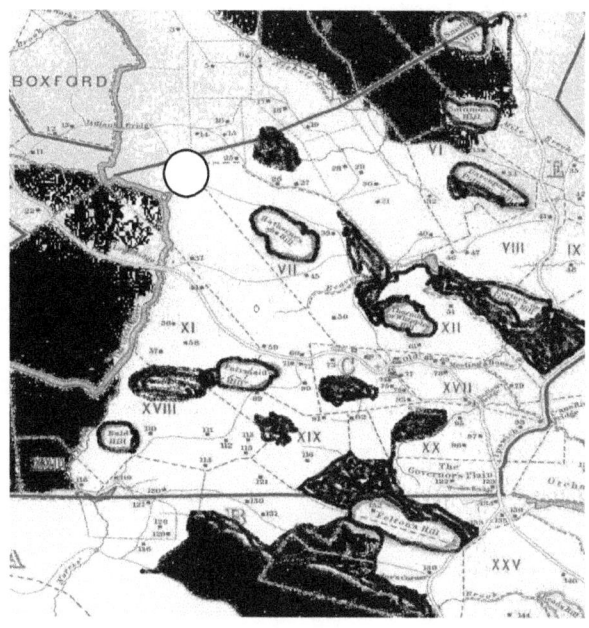

**MAP 11:** places some known 17th C. landscape features onto the Upham map to create a depiction of the land in 1692. The area included the large Hathorne Meadow area. The Thomas and Edward Putnam families had houses located in the area under the circle, their cousin John Putnam Jr. lived nearby. Black areas are areas where trees were either described as growing, or where the soil map indicates that there are steep stony hillsides that could not be cultivated. The Rea Farm was situated upon Hathorne Hill. These families included the afflicted Ann Putnam, Sr. and Ann Putnam, Jr., several of the afflicted girls and Hannah Putnam. They all suffered from neurological "fits." Ann Putnam, Sr. may have suffered a miscarriage and the eight-month-old infant of John Jr. and Hannah died after exhibiting the same symptoms. Edward Putnam described what may be an ECM on the skin of one of the afflicted. An overlay was created based on written descriptions of part of the property, including specific oak marker trees, from the 1685 will of Thomas Putnam, Sr. There may be more forested spots or areas, especially in the north towards Boxford. This map shows what could be a **HIGH RISK** level of Lyme disease risk because of the high level of edge interspersion between herbaceous zones and forest. The landscape includes numerous rocky hill edge ecotones,

 DISGUISED AS THE DEVIL

forest fragments, orchards, a riparian corridor and swampy wet area that would attract deer and other animals in a time of drought. Numerous oak trees were mentioned. If pigs were removed from this landscape and/or the burning off of undergrowth and leaves were neglected or not done, it would be an even greater "geography of risk."[521]

---

**SALEM VILLAGE**

# SALEM VILLAGE

[492] Eugene Stratton, *Plymouth Colony* (Salt Lake City: Ancestry Publishing, 1986), 50.
[493] Virginia DeJohn Anderson, *New England's Generation: The Great Migration and the Formation of Society in the Seventeenth Century* (New York: Cambridge University Press, 1991), 15-17.
[494] Daniel Gookin, *Indians*, 7-12.
[495] William Wood, *New England's Prospect*, 62.
[496] D. Cressy, *Coming Over*, 275-276.
[497] Mark Lapping, Ed., Howard S. Russell, The *Long Deep Furrow: Three Centuries of Farming in New England* (Hanover, N.H.: University Press of New England, 1982), 96-97.
[498] John Smith quoted in James Duncan Phillips, *Salem in the Seventeenth Century* (New York: Houghton Mifflin, 1933), 15.
[499] *Ibid.*, 71.
[500] Frances Rose-Troup, *Roger Conant and the Early Settlements on the North Shore of Massachusetts* (n.p.: Roger Conant Family Association, 1926), 11-12.
[501] Francis Higginson quoted in Emerson, Ed. *Letters*, September, 1629.
[502] See Sidney Perley, *History of Salem* (1924), Map of Salem in 1700 and Charles Upham, *Salem Witchcraft*, (New York: Frederick Ungar Publishing, 1867), 34-35,144-145.
[503] Phillips, *Salem*, 71, Upham, *Witchcraft, Vol. 1*, 31-33.
[504] Boyer and Nissenbaum, *Salem-Village Witchcraft*, 236.
[505] Phillips, *Salem*, 111-114.
[506] Boyer and Nissenbaum, *Salem-Village Witchcraft*, 230-231.
[507] Upham, *Salem Witchcraft*, Vol. 1, 147.
[508] Upham, *Salem Witchcraft*, Map of Salem Village in 1692.
[509] *Ibid.*,31.
[510] Sarah Loring., *History of Andover* (1880).
[511] Lapping, Ed. *Furrow*, 96-97.
[512] Boyer and Nissenbaum, *Witchcraft*, 315.
[513] Upham, *Map of Salem Village in 1692*.
[514] Frances Hill, *Hunting for Witches* (Beverly, Massachusetts: Commonwealth editions, 2002), 86-133.
[515] John Winthrop, Jr., *Records*, n.p. Massachusetts Historical Society.
[516] Nathaniel Shurtleff, *Records*, June 14, 1642.
[517] Boyer and Nissenbaum, *Witchcraft*, 236-237.
[518] Upham, *Map of Salem 1692* and The Andover Historical Society, *Map of Andover, 1692*.
[519] David Hackett Fischer, *Albion's Seed: Four British Folkways in America* (New York: Oxford University Press,1989),183-84. The village split over the witch accusations involves a blend of East Anglian and West Countrymen that is dominated by geographic location in 1692 more than area of English origin. The Putnams were East Anglian, the Porters were probably West Countrymen but Bray Wilkins, an ally of the Putnams was possibly of Welsh descent and married to a West Country woman. He lived near the Putnams in the western part of Salem Village.
[520] This map is based upon the concepts of Native American inhabitation discussed by Brian Donahue in *The Great Meadow: Farmers and the Land in Colonial Concord*. New Haven: Yale University Press, 2007, which describes land use practices in Concord, which is located west of the Salem Area. Underlying map: United States Department of Agriculture, Soil Conservation Service, *Soil Survey of Essex County: the southern part*, (September 1986).
[521] Underlying map: Upham, *Map of Salem 1692* (1867).

# DISGUISED AS THE DEVIL

POLITICS AND PLACE

# IX. WHAT HAPPENED IN SALEM, 1692? PLACE, POLITICS AND SOCIAL CONTEXT

It has been noted that there is often "a monumental discrepancy between what men think they are designing and the world they are in fact building."[522] This may be particularly true of the early Roman Catholic Church, in the historic case of the Puritans of Massachusetts, and is still relevant today. The reactions to the afflictions that resulted in the witch trials of 1692 took place within a specific social context. The Massachusetts Bay Colony had been founded by a group of English citizens with a strong ethic revolving around a relatively newly formulated set of protestant principles called Puritanism. The world that they began to build in the 1630's ended in ways that they would never have conceived. As the colony grew and matured it faced the inevitable aging and decline of its founding generation, an internal crisis in the 1660's about the requirement for a covenant with god for civic/church membership, and the eventual immergence of new set of second generational leaders. Looking at this civic society, it can be argued that what began as a centrally located, economically flexible but ideologically rigid, entrepreneurial endeavor in 1628 had by 1660 evolved into a more geographically and structurally disbursed organization. It had developed expanded markets for its goods and many civic responsibilities had been delegated to the local church/town level.

      Emile Durkeim stated that a society presents its distinctly corporate capacity through a variety of symbolizations or collective representations. A collective society realizes its corporate existence through a structural apparatus that is empowered to make collective decisions and take collective action. Durkheim felt that most collective representations are created or renewed by the definitions used in a sacred/profane dichotomy. For example, a societal process might be created in response to both religion and crime. The degree to which these collective symbolizations permeate daily life is described by the term 'immanence.'[523]

Although a rationalist, Max Weber's discussion of the historic transition between the two ways of thinking is relevant to any discussion of the cultural environment of the seventeenth century. Western society, including the Puritans in Massachusetts, came to be in a state of transition between a traditional set of beliefs and an emergent rationalist or scientific model. This transition required what has been previously noted as "the disenchantment of the world" which entailed belief in the primacy of science being allowed to flow into the society's normative cognitive system. The outflow would contain the enchanted belief in "mysterious incalculable forces" that were shaken loose by the rejection of superstition.[524] Science became a god, some would say

Historically, corporate organizations have redefined themselves periodically by denouncing those who allegedly violate cherished collective representations. The happened in Europe when the Roman Catholic Church responded to the threat of heresy with an Inquisition, which later melded into the great Age of European Witch Hunts. In *The Sacred and the Subversive: Political Witch Hunts as National Rituals,* Albert Bergesen looked at the political organizational structures of one party, two party and three party systems. He found that "in more highly immanent societies there will be discovered a larger volume of deviance in more institutional sectors than in less immanent societies." He also found that "ritually created subversion is more dispersed throughout the social structure" in one party states.[525] The organization of the Early Modern Roman Catholic Church, the seventeenth century Massachusetts Bay Colony, and some modern medical associations can be described as one party theocracies with a highly immanent collective organization.

When King James I issued *The Charter of The Massachusetts Bay Company* in 1629 it was as remarkable for what it left out as well as what it included. It granted all lands between "the northerly part of the colony of Plymouth to the south, and the Piscataqua River to the north being commonly known as Massachusetts Bay." The Atlantic Ocean bordered on the east but there was no western boundary mentioned.[This would allow for the expansion of this colony into the territory called Connecticut] The document is, in general, very concerned with trade and

## POLITICS AND PLACE

economic opportunities and very liberal in the delegation of power and distribution of civic authority. It states that the company shall have "a governor, deputy governor, and assistants." They should "keepe a court or Assemblie of themselves, for the better ordering and directing of their affairs." It leaves the creation of rules and laws for governing the colony in the hands of the colonists themselves while assuring them that both they and their offspring would retain the right of English citizenship. The company is given "full and absolute power and authority to correct, punishe, pardon, governe and rule" the subjects living under their jurisdiction.[526] This charter allowed the Massachusetts Bay Company to function as a theocracy with minimal supervision from the English government for over fifty years. When this charter was revoked in 1686, in triggered a chain reaction of crisis in government, leadership, and, ultimately, was connected to the witchcraft persecutions in New England.

John Winthrop's *A Model of Christian Charity* Sermon was written during his transatlantic crossing from England to the new Massachusetts colony. This moral justification for the Puritan "errand into the wilderness" is an important document because it delineates the idealism and the philosophical entrepreneur-like spirit of this founder of the Massachusetts Bay Colony. This sermon has been widely quoted, perhaps most famously by President Ronald Reagan. It models the founding of America as "a covenant" with God for his work that "may be emulated by the world…we must consider that we shall be as a city upon a hill, the eyes of all people are upon us" and if successful "men will say of succeeding plantations" please " lord, make it like that of New England." [527]

John Cotton's *Democracy as a Detriment to church and state* uses the Socratic device of demands or questions and answers to justify the establishment of this theocracy as the best form of government for New England. It identifies two classes of people-the nobility and the people who are allowed hereditary liberty as freemen. However, the ordinary practice of the Massachusetts commonwealth is described as being "that none are admitted freemen but such as are first admitted members of a church." John Cotton found this to be compatible "for the liberties of the freeman of this commonwealth are such, as require men of faithful integrity to God and the state, to preserve the fame." The direction of God made a "theocracy in both, as the best forme of government in the commonwealth, as well as in the church." The government as set up was described as "not a democracy" although the people choose their leaders.

 DISGUISED AS THE DEVIL

Cotton writes that "nor need we fear that this course will, in time, fast the commonwealth into distractions, and popular confusions" for these three things "doe not undermine, but do mutually and strongly mainteyne each other:….authority in magistrates, liberty in people, and purity in church. Purity preserved in the church, will preserve well ordered liberty in the people and both of them establish well-balanced authority in the magistrates." The covenant with god required for church membership was inextricably linked in John Cotton's mind with a covenant with the civic organization. [528]This covenant included an obligation to support the church financially.

In *Ordering Their Private World,* Charles Hambrick-Stowe offers a good description of the immanent nature of the Massachusetts Bay society. Puritanism incorporated what he calls "the inner life of the soul" into daily life through both public worship and especially private devotion. Puritans meditated and prayed just before going to sleep at night, upon rising in the morning, on Saturday in preparation for the Sabbath, on the Sabbath between the two regularly scheduled three-hour long services. These services were held in a multifunctional meetinghouse that was also used for civic meetings and court sessions. Special sessions of meditative self-examination could be prompted by a birthday, the beginning of a new year, or after some remarkable "providence" in one's life. They participated in private conferences with ministers and religious leaders. Families had devotion services, sometimes several times a day and during the week there were neighborhood prayer meetings held in private homes. Diaries were used to record spiritual experiences. Religious publications-sermons, tracts, catechisms and devotional manuals were best sellers. The New England Primer for children began with A is for "In Adam's fall we sinned all." Puritanism permeated all aspects of the Massachusetts culture, including economic activities.[529]

Cotton Mather is a good example of an emergent second generational leader of the Massachusetts Bay Colony who became entangled in the witchcraft trials. His Diary and biography illustrates both the concept of the immanence of Puritanism and the challenges that were faced in the colony during Mather's lifetime. Mather is a good example of the leadership of early Massachusetts-educated at Harvard to be a theologian; Mather also participated in formal civic government organizations, the militia, and served in some judiciary roles. He was also an active participant in the informal social life of the colony. He socialized with a set of individuals who were often members of the same formal organizations. In his diary, the pressing matters of the church and colonial government spill easily between the two spheres. Matters discussed at dinner or in a private letter often appear up in the official records of the formal organizations of the colony at the same time.

Mather was also involved in the importation of scientific knowledge from Europe that helped open the gateways for the inflow of scientific knowledge into Puritan society. He was extremely proud of the honorary doctorate that he received. The boundary crisis that was created by the scientific redefinition of various phenomena played a role in his life. At almost the same time that he was decrying witchcraft and supporting its designation as a capital crime, he was laying a foundation for scientific research.

Mather's relationship with God and the extent that religion permeated his life is apparent in his diary entries. Groveling face down on the dusty floor of his study was a daily occurrence. Both the everyday and the momentous were dissected in his mind for interpretations of the will of God. Unsettling at first to the modern mind, the saturation and surety of religion seems to have provided an anchor for this puritan's life while also creating a source of some discomfort. Mather sat at the brink of the eighteenth century's scientific revolution. He was mentally challenged and actively involved in the furtherance of medical knowledge but his science was deeply permeated by what he saw as the hand and will of God.[530]

The transition between the founding generation in New England, made up of the group of early settlers who migrated as part of the puritan and separatist movement out of England in the early seventeenth century, and their descendents was a difficult one. In a theocratic society where church membership also defined civic rights and responsibilities, a crisis evolved over the personal "covenant" between individual and god that was required for church membership. It became a problem when the "first generation" of Puritan offspring failed to develop the same religious fervor that their parents had exhibited and often failed to experience the "covenant." Without the covenant, a person could not become a member of a church. The halfway covenant compromise, allowing church membership to the children of people who were already church members solved the problem for a while but was still being debated as late as 1692. There was a generational gap between those who had suffered persecution in England and who had made the difficult transition to the New World and those who had simply been born into it. The founding

## DISGUISED AS THE DEVIL

generation resisted change but the next generation was forced to embrace it when the English government revoked the colony's charter. The original charter was replaced by a new one that denied civic rule limited to a single religion and expanded suffrage to among others, Quakers, who had been heavily persecuted in Massachusetts. The new charter expanded the new Massachusetts Commonwealth's territory by annexing it with the Plymouth Colony, a move that was also met with some resistance.[531]

The witch-hunt in Salem Village can be looked at as part of a crisis that involved both mental constructs and a geographical edge environment. The collective structure of Salem Village consisted of a social network that has been well analyzed as being divided locally on an east/west dichotomy. The Putnam family and their allies created a power cluster in the west and the Proctor family were the leading and most prosperous family in the eastern section of the village. A set of informal, familial, economic and formal social networks and organizations can be delineated.

The witch- hunt has been seen as an outgrowth of conflicts between the Proctors, as a rising merchant class, and the Putnam's land based economy. Two interesting points to note are the importance of the church's meeting house as the central stage for many of the events that unfolded, underscoring the imminence of religion in Puritan civic life, and the multi layered lives led by many of the residents of Salem. From a feminist standpoint, women play a role in this story that is more diverse than might be supposed. Most of the accused witches and their accusers are female, and women also testify, play a role as medical experts and witch mark examiners, and have formal full church memberships in Salem Village.[532]

By 1692, it must have seemed like everything was falling apart at the seams. Far from being a model of a shining "city on a hill," in 1692 Salem Village was a town under stress in a colony that was bankrupt. It had suffered from a prolonged period of drought. War with heathen enemies that worshiped the devil himself was raging all around. Death and the devil came from the forest. The territory of Maine was a burned out husk that was in the process of being abandoned by their Massachusetts Bay government. A stream of traumatized refugees had straggled down from the Eastward, as Maine was called, seeking refuge, and many were housed in Salem. Even the sun, blurred by the smoke of numerous fires set in the Eastward, must not have seemed to be as bright as it once was.

## POLITICS AND PLACE

In the late summer of 1690 most of the Massachusetts Bay Colony was "in armes" and when less than a quarter of the 3000 men needed volunteered, men were drafted or pressed into service. By August 9, 1690, the colony had organized and supplied an invasion force that was sent to Quebec. Members of the militia, organized by town, were extracted at a crucial time of year-right before the harvest, causing a critical labor shortage. While it did recover some hostages, the Quebec invasion was a disaster. As many as one third of the militia members died, mostly from smallpox. Many ships were lost and men drowned. [533] Some returning veterans were "walking wounded"- they had been affected both physically and mentally. Several testified during the trials of 1692. The government of the colony was bankrupt, owing more to the merchants that had financed the invasion that it was able to raise in taxes. As a final insult, there was no money to pay the militia after it returned in defeat. Governor William Phips was forced to use some of his personal funds to stave off public riots.

William Phips was the first governor for the Massachusetts Commonwealth to be appointed by the new English King William under the new charter of 1692. Born in Maine, Phips' adventurous life had led him to salvage a sunken Spanish treasury ship in the Caribbean. When he paid the British government the legally required 'portion' of the gold and silver that he had raised from the sea floor, it can be argued that his treasure hunting activities had enriched the English treasury with almost enough money to finance Parliament's Glorious Revolution. It was also enough for him to be awarded with a knighthood. Phips led the 1690 assault on Quebec, was in office during the Salem Witch trials, and eventually called for their official halt. The combination of King Williams War, a new chartered form of government, and the drought created a major crisis for the colony that Phips had difficulty managing.[534] The outbreak of disease in Salem Village could only contribute to the colony's, if not England's, ongoing mental state of crisis. It was a time of change.

Immanuel Kant, pondering the Scientific Revolution that occurred during the sixteenth and seventeenth century in Europe wrote that in the seventeenth century "a new light flashed upon all students of nature… [that study] entered on the secure methods of a science after having for many centuries done nothing but grapple in the dark." This is a fairly accurate description of what was to be a revolutionary change in mindset that occurred during this time period. Some theorists note that the fundamental Puritan requirement of a first hand experiential covenant

DISGUISED AS THE DEVIL

with God was not dissimilar from the personal observation and experimentation that were the hallmarks of scientific research. The Protestant Revolution may have paved the way for proto modern attitudes by setting the precedent of the rejection of the long standing prevailing belief set [of the Roman Catholic Church] and replacing it with newly formulated ones.

At a time when religious questioning was at a peak, there was a growing skepticism about certainties that were founded upon faith rather than reason. The scientific desire for rational explanations led to the ultimate rejection of belief in miracles, witchcraft, or other explanations for experiences that could not be explained in scientific terms.[535] On a philosophical level, however, the conflict would be long and drawn out with remnants of earlier belief systems remaining stubbornly persistent in popular culture even to the modern day. Witches were accused well into the eighteenth century in both Europe and America and Santa Claus still uses his entry chimney once a year.

The Salem Witch trials testimony and the Witchcraft Reparations of 1704 in the Massachusetts Bay Colony, however, show a diversity driven attitudinal change occurring within a very short time frame in New England. Looking at land use and architectural patterns as emblematic symbols of the social paradigms that created them, the remnants and records of this time period also show this shift. The earliest settlers organized in what can be described as an organic medieval way that changed rapidly during the last ten years of the seventeenth and early eighteenth centuries. After about 1700, housing designs, plot plans, and land use patterns acquired a new balanced and orderly style that is often called the "Georgian Mindset" in the English influenced world. The changes that occurred in the objects and layout of regular daily life show that, once it took hold, science became a powerful and culturally immanent force.[536]

In *The Disenchantment of Magic: Spells, Charms, and Superstition in Early European Witchcraft Literature,* Michael D. Bailey writes that one of the hallmark features of modern western thought that was born of the protestant reformation but which came into vogue during the Enlightenment is what Max Weber called "the disenchantment of the world." This entails a belief in the primacy of science and that "there are no mysterious incalculable forces" in the world. Issues of "magical thought" and "superstition" in opposition to "scientific rationalism" frame many discussions in which western modernity is juxtaposed against the

POLITICS AND PLACE

traditional beliefs and practices of the historic past.[537] Salem Village of 1692 sat in the middle of this clash.

When the new worldview of rational scientific experimentation and knowledge began to flow into the Massachusetts Bay collective organization it took time for many features of earlier thinking to be transformed. The crisis of instability and fear that the Salem Witch trials emerged from would be remembered, even as the concept of witchcraft had been transformed into a cardboard cutout for Halloween decorating by the end of the eighteenth century. "We shall see the reign of witches pass over, their spells dissolve, and the people, recovering their true sight," Thomas Jefferson posited,[538] as rationality permeated the world. However, there is a resilience and power in seventeenth-century popular culture that still permeates American history. The colonial climate of fear and inner instability continued to have an impact long into the next century. The excesses of the Puritan Experiment are often cited as the motive behind the United States Constitution's separation of church and state, the Fifth Amendment (one accused witch who refused to plead or testify was pressed to death under a pile of rocks in 1692), compensation for government confiscation of private property, and various aspects of the American Revolution.[539] It still makes some question the compassionate Christian conservative movement's motives.

    Kai T. Erikson applies Durkheim's theories about deviant behavior, boundary issues, and immanence to the Puritan society of 17th-century Massachusetts. He uses the records of the Massachusetts Bay Colony to illustrate the way in which deviant behavior fit into the texture of collective social life and how deviant forms of behavior are often a valuable resource in society. Deviance provides a point of contrast, which is necessary for the maintenance of a coherent social order. The ritual and symbolic dimensions of social life play an important role in the creation and maintenance of any collective boundary. He also identifies boundaries as a crisis ridden but critical organizational phenomenon. By creating and defending an almost impenetrable social boundary the Massachusetts Bay Colony also created its own downfall as a mono-theocratic entity.[540]

    The early Roman Catholic Church can also be analyzed in a similar way. By defending its power within an inflexible collective boundary, defining deviance in dramatic and prurient terms in contrast to the salvation associated with its dogma, it attempted to maintain a pre-existing social order. What it got instead was bloodshed, torture, the disorder of the Protestant Reformation and a loss of power.

## DISGUISED AS THE DEVIL

The power structure in New England underwent a change in 1692. During the 1680's the English government had received so many complaints about the Massachusetts Bay Colony that the King pulled its charter. John Winthrop's "city on a hill" had included no mechanism for dealing with dissent other than corporal punishment, exile, and in some cases the death penalty. Quakers were attacked and even the King's officially sanctioned Anglican Church was discouraged from setting up shop during the restoration period. The responsive 1692 *Charter for the Commonwealth of Massachusetts* radically changed the civic government of the colony forever.

This document included the provision that the electorate would no longer be limited to Puritans (the covenanted) but broadened to include propertied members of every Christian sect (except Catholics). This was revolutionary because the non-covenanted population far outnumbered those within it and it meant legally sanctioned acceptance for the particularly despised Quakers. This gave the power of the majority to non-Puritans. The new charter also changed the governor's office from being one that was elected within the colony to one that was appointed by the King of England.[541] Within a short period of time after the imposition of the 1692 charter, the witch trials were ended and within a few years, the colonial government voted to financially compensate the victims of the Salem Witch Hunt. Many official apologies have followed including one issued by the governor of the Commonwealth of Massachusetts in 1992.[542]

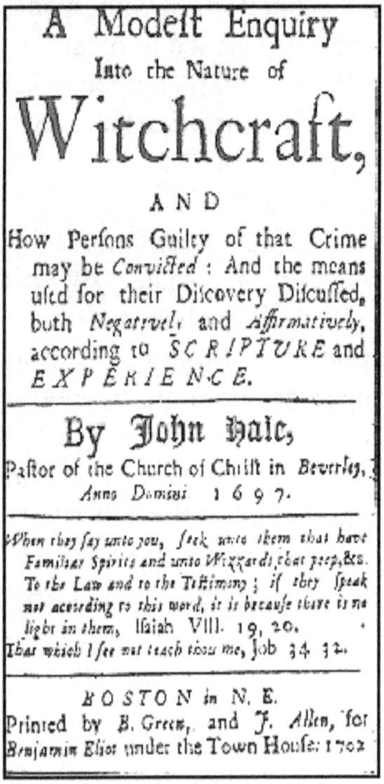

A Modeſt Enquiry
Into the Nature of
Witchcraft,
AND
How Perſons Guilty of that Crime may be *Convicted*: And the means uſed for their Diſcovery Diſcuſſed, both *Negatively* and *Affirmatively*, according to *SCRIPTURE* and *EXPERIENCE*.

By John Hale,
Paſtor of the Church of Chriſt in *Beverly*,
*Anno Domini* 1697.

*When they ſay unto you, ſeek unto them that have Familiar Spirits and unto Wizzards that peep, &c. To the Law and to the Teſtimony; if they ſpeak not according to this word, it is becauſe there is no light in them,* Iſaiah VIII. 19, 20.
*That which I ſee not teach thou me,* Job 34. 32.

*BOSTON* in N. E.
Printed by *B. Green*, and *J. Allen*, for *Benjamin Eliot* under the Town Houſe: 1702

Figure 43. *Frontispiece, 1702.*

In one party systems, like the early Roman Catholic Church and the theocracy of early Massachusetts, the party represents the interests of the state as a whole, and there are no formal arrangements for sub-state level groups to have their constituent interests formally represented. It was not even possible to "agree to disagree." The same thing appears to be happening with modern Lyme disease treatment protocols. One party line has been given preference above all others in the United States by the CDC and there is no formal platform for dissenters to express their opinions to the contrary. This structure is backed by a firm belief in the "theology" of the twentieth century: Science. This establishment has countered any criticism of its interpretation of the science of Lyme disease by formulating a set of demeaning labels: Quacks, the Lyme cult, the Lyme counterculture, an imaginary affliction, and applying them liberally to the powerless. They use the formal disciplinary procedures available to them to chastise doctors who treat people who appear to suffer from the chronic form of Lyme disease. Once again, it is not possible to even agree to disagree, and woe to those who do.

Using Shakespeare's "all the world's a play" as an analogy, we can begin to look at events in history as a complex web of interlocking components, much like a production of Hamlet in the Park. Politics, social belief systems, genetics and human personalities wrote the script, provided the players, created very specific costumes and makeup, provided the language, and supplied an empathetic and understanding contemporary audience. All of these components of the 1692 witch hysteria are important. But the setting, the stage, and the all-important backdrop that land use and the environment provided for this story are also important. It is the hypothesis of this text that a unique but historically ubiquitous convergence of environmental factors created the affliction that was the impetus for the witch trials of 1692 and the modern "Lyme controversy." How the Puritan society handled affliction is what makes it stand out in history. English cultural practices were carried out as a part of day-to-day life within the context of political, economic, and social upheaval and change.

The people of Salem Village lived in a colony at war "to the Eastward," and suffering from economic instability and privation. They were a colony with a new charter that had been negotiating with a mother country in the middle of a civil war. While historians have turned this into the Glorious Revolution, it was a bloody fight that pitted parliament and their newly created King William and Queen Mary against the deposed King James II. The truly deciding Battle of the Boyne was not fought in Ireland until 1690 at a time when warfare with the French and Native Americans was already in full swing in New England.[543]

In the late summer of 1690, the already drought stressed agricultural areas of the Massachusetts colony were given short notice to supply an invasion force that would be sent to Quebec with men and

## DISGUISED AS THE DEVIL

foodstuffs. The environment in Salem Village would be affected by this action for several years in the future. Members of the militia, organized by town but crucial as an agricultural labor force, were extracted at an important time of year: right before harvest. Cattle, consisting of both pigs and beef stock would have been herded onto the waiting invasion fleet's ships or slaughtered and casked up in briny water for later use. There was also the omnipresent problem of marauding Natives. In the Kittery, Maine area the "enemie" was "constantly killing and destroying bothe fatt and lean cattle and it is taken for granted without some speedy help com that they will not leave a beast alive in the whole province."[544]

This left the woodlands surrounding Salem Village and elsewhere virtually empty of pigs at an ecologically sensitive time of year-mast season. Shorthanded farmers would have been hard pressed to even collect many acorns for the breeding stock that they would have retained. However, for some wildlife residents of the forest in the autumn of 1690, especially white footed mice, this would have been a mast bonanza year-similar to the conditions created when the modern scientists augmented with acorns at their test sites. More mice and white tailed deer than normal may have been attracted to the suddenly abundant mast, bringing a larger than normal population of ticks closer to the homes of the settlers than ever before. The fact that the colony was in the middle of a prolonged drought would have also influenced this wildlife population-moving them especially toward water sources like the Ipswich and Shawshin Rivers and Five Mile Pond. This dynamic would have begun as early as 1681 and continued until late 1692 when the drought finally broke.

When the planned "quick and easy" invasion of Quebec turned into an ordeal that some militiamen did not survive [we might call it a cragmire], the area in the proximity of English colonial settlements was probably also given a reprieve from another important tick control measure-burning. Typically done in the winter or early spring months, burning off the ground was a male chore that may have been neglected for the years 1690 and 1691 if it was done at all during the drought. Firewood was in short supply, and even the village minister seems to have to beg for firewood or for a decent communion table to be supplied with bread. Food supplies must have been affected by the seemingly endless drought. In addition, there was simply not enough labor available to get everything done.

The mast season of late summer and autumn 1690 may have caused the symptoms of acorn poisoning in cows and horses that are recorded in the Salem Witchcraft Trial testimony, although the date of these events is unclear. The residents of Salem Village were used to pasturing horses and cattle "in the woods" but this is a practice that only works well when the woods are also inhabited by acorn consuming pigs. With the pigs out of the woods and on their way to Quebec, it became problematic. Horses are vulnerable to ingesting the acorns that are found in their woodland pastures. It appears that several became sick or died. Thomas Andrews testified that a friend's " mare was not well …her lips ware exceedingly swelled that the insides of them turned outward and looked black and blew and gelled her tongue was in the same condition…"[545] This is a good description of a horse suffering from acorn poisoning. Another mare mentioned with severe colic and famously "exploding farts" may have been another victim of this malady.

The high and dynamic population of white footed mice that a mast year creates, combined with a drought induced concentration in areas near fresh water in the western side of Salem Village, a larger than usual population of nymphal and adult ticks that they nourished, infected, and moved around between forest and patchy areas, would have made the nearby human population more vulnerable to Lyme disease than usual. This appears to be what happened. By late 1691, an increased number of afflicted humans began to be recorded. Many of them were young, skirt wearing (and dragging) females. In addition, infection continued beyond 1692 even after the witch trials had ended. It continues in the present day in Massachusetts and elsewhere.

Graphical analysis of those affected by witchcraft as either an accused witch or as afflicted as described in seventeenth century literature shows a bimodal age distribution that is similar to the chart for modern Lyme disease patients [See Charts B]. The young were afflicted, the old were accused. Marks on the skin are mentioned: Devil's marks and witches teats were searched for and found on the bodies of accused witches. Four-year-old accused witch Dorcas Good, for example, developed "a deep red spot about the biggness of a fleabite" on her finger. But a sizable number of the afflicted also developed marks on their bodies that were, in the majority of cases, noticed and often described as looking like bite marks. In June of 1682, the second year of the drought, Mary Hortado of Salmon Falls in Maine "was bitten" leaving "the impressions of the teeth being like a mans teeth" that "were plainly seen by many." Jarvis Ring testified that he had the "print of a bite of a woman" on his right hand. The Goodwin children developed red streaks on their bodies. In 1692, the afflicted Mary Wolcut and others had the

## DISGUISED AS THE DEVIL

marks of "a small set of teeth." Deodot Lawson described these marks as being made by teeth "both upper and lower set." The afflicted Abigail Williams developed a mark "like the exact print, image and color of an orange" on her leg. Margaret Rule had blisters that developed 'like common burning' that would be cured in two or three days at the furthest. Mercy Short also developed blisters (like a scald) that went away in a day or two.[546] These prints may have been the rashes associated with Lyme disease and its co-infections.

After suffering from affliction, a few of the afflicted died, but most got over it and then went on to live their lives in relative obscurity. The four-year-old accused witch Dorcas Good and Betty Parris' brother Noyes were mentally ill for the remainder of their lives. Some, like Ann Putnam Jr., continued to suffer from a chronic form of the disease for the rest of their lives. Others, like the Godwin children, seem to have become fully cured. In addition, the infection may have made one earlier sufferer, Elizabeth Knapp, fat- perhaps with a form of what has been described as infect-obesity.[547] All in all, the incident was remembered as more of a social phenomenon than a lasting ordeal. Past land use practices along with devil's marks and the belief in witchcraft have all been obscured by both time and changes in cultural mindsets.

As Salem Village developed, the fragmented land use pattern of 1692 made way for the well-cultivated landscape of late eighteenth century New England. Forest cover was removed and replaced by herbaceous cover in the form of grass and crops. Industrialized areas were added. Deer hunting became an important component of northern New England's culture. Domesticated animals grazed the land. This created an environment that was not conducive to Lyme disease. It became a disease of the forest edge but that edge was usually far away from human habitation. Lyme disease remained an unnamed and extremely infrequent affliction. That changed over time. The lure of fertile, and less rocky, farmland in the Midwest depopulated many parts of New England during the 1820's and 30's and led to reforestation in those areas. The Industrial Revolution also changed the landscape of New England by concentrating the population in urban areas for the first time. The forests began to grow back and eventually the deer came back too. With them came ticks. This came at a time in the nineteenth century when wealthy individuals were building homes in country places to get away from an increasingly unhealthy urban environment. They were also interacting with nature in what was becoming a popular leisure activity called "rusticating." It was the age of the great resort hotels in the mountains. Not surprisingly, a

## POLITICS AND PLACE

"new" disease then began to emerge that was linked to these upper class activities: neurasthenia.

Writing in 1869, George Miller Beard was the first to label this "new" disease.[548] Symptoms included fatigue, anxiety, headache, impotence, neuralgia and depression. It was explained as being a result of exhaustion of the central nervous system's energy reserves, which Beard attributed to the stresses of civilization and urbanization. It was associated with upper class individuals in sedentary employment. Women [who were still wearing long dresses] were said to be particularly prone to the condition. People suffering from neurasthenia would find themselves suddenly unable to function, stymied by fatigue, weakness, strange pains, dizziness and passing out. In men, it caused erectile dysfunction. Doctors could not find anything to explain these symptoms, so they attributed them to a "weak nervous system." Women with neurasthenia were sent to the country for rest cures or confined to their beds, where they would either recover or, eventually, die.[549] Science never found a causative agent for neurasthenia and accounts of it tended to be washed away by the great influenza pandemic that began in 1918.

Animal pelts from this time period that have been preserved in museums have been found to be infected with *Bb,* so it is possible to also propose Lyme disease as a causative agent for some of these historic victims of neurasthenia. This disease entered the twentieth century as an affliction acquired during interactions with the forest edge. It affected more women because they were still wearing their long tick collecting skirts during the early part of the century. And even more frequent interactions with wooded edge areas began in the late twentieth century when the American middle class moved out into the post war dispersed settlement pattern of what had become prime Lyme disease habitat: the suburbs. The old sickness returned. As had happened many times before, population movements accompanied by deforestation created the modern chapters of the very long and very old Lyme disease story that we live with today. How modern society responds will end up as the measure of our mettle.

## DISGUISED AS THE DEVIL

# WHAT HAPPENED IN 1692?

522 Scott, W.R., *Organizations: Rational, Natural and Open Systems* (Upper Saddle River, N.J.: Prentice Hall, 2003), 185-190.
523 Emile Durkheim, *The Elementary Form of the Religious Life* (Oxford University Press: New York, 2001).
524 Michael D. Bailey, "The Disenchantment of Magic: Spells, Charms, and Superstition in Early European Witchcraft Literature," in *American Historical Review*, Vol. III No.2 April 2006, available online at www.historycooperative.org/journals/ahr/111.2/bailey.html.
525 Albert Bergesen, "The Sacred and the Subversive: Political Witch Hunts as National Rituals" *Monograph Number 4* (Storrs, Ct: The Society for the Scientific Study of Religion, 1984).
526 King James I *The Charter of The Massachusetts Bay Company* (1629)
527 Winthrop, J., *A Model of Christian Charity Sermon* (1630) in Essential Documents in American History at Academic Search Premier http://ebscohost.com.
528 Cotton, John. *Democracy as a Detriment to Church and State* (1636) in Essential Documents in American History at Academic search Premier http://ebscohost.com.
529 Charles Hambrick-Stowe, "Ordering Their Private World," *Christianity Today* International Issue 41, Vol XIII, No.1 May 1996, 16.
530 In Silverman, K., *The Life and Times of Cotton Mather* (New York: Columbia University Press, 1971).
531 Virginia D. Anderson, *New England's Generation: The Great Migration and the Formation of Society and Culture in the Seventeenth Century* (U.K.: Cambridge University Press, 1991).
532 Boyer and Nissenbaum, Salem.
533 Marc-André Bernier, "The 1995 Survey of a Ship from Sir William Phips' Fleet, 1690," *Underwater Archaeology* (1997).
534 Baker, E. W. and J. Reid, *The New England Knight: Sir William Phips, 1651-1695 (*Toronto: University of Toronto Press, 1998).
535 Smith, A. G. R. *Science and Society in the Sixteenth and Seventeenth Century*. (Norwich, U.K.: Jarrold & Sons, 1972).
536 Robert St. George argues this in *Conversing by Signs: Poetics of Implication in Colonial New England Culture* (Chapel Hill, N.C.: The University of North Carolina Press, 1998).
537 Bailey, *Disenchantment*.
538 Thomas Jefferson, writing about the Aliens and Sedition Act in 1798, from *The Thomas Jefferson Papers, 1606-1827*, original manuscript documents at the Library of Congress online collection at http:memory.loc.gov.
539 The Fifth Amendment reads: No person shall be held to answer for a capital, or otherwise infamous crime, unless on a presentment or indictment of a Grand Jury, except in cases arising in the land or naval forces, or in the Militia, when in actual service in time of War or public danger; nor shall any person be subject for the same offense to be twice put in jeopardy of life or limb; nor shall be compelled in any criminal case to be a witness against himself, nor be deprived of life, liberty, or property, without due process of law; nor shall private property be taken for public use without just compensation.
540 Erikson, K. T., *Wayward Puritans: A Study in the Sociology of Deviance* (New York: Wiley, 2004 Reprint).
541 King William and Queen Mary, *The Massachusetts Bay Charter of 1692* in Essential Documents in American History at Academic Search Premier http://ebscohost.com.
542 Chapter 122 of the *Acts of 2001*, Commonwealth of Massachusetts (see http://www.mass.gov/legis/laws/seslaw01/sl010122.htm); "New Law Exonerates", *Boston Globe* (Nov. 1, 2001).
543 For a good description of this battle and the war, see Padraig Lenihan, *1690 Battle of the Boyne* (Gloucestershire: Tempus Publishing, 2003).
544 Mary Beth Norton, *Snare*, 109-110.
545 Thomas Andrews testimony, Essex County Archives, *Salem Witchcraft* Vol. 1,152
546 Either a lot of biting was going on or maybe these people were developing ECM's.
547 M. Pasarica, and Dhurandhar NV. "Infectobesity: obesity of infectious origin," *Adv Food Nutr Res*, 52(2007), 61-102. See also Robin Marantz Henig, "Fat Factors" *New York Times Magazine* (August 13, 2006).
548 George Beard's article was "Neurasthenia, or nervous exhaustion" *The Boston Medical and Surgical Journal (*Apr. 28, 1869), 217-221)
549 Francis Gosling writes about this in *Before Freud: Neurasthenia and the American medical community, 1870-1910* (Urbana: University of Illinois Press, 1987).

CONCLUSIONS

# X. CONCLUSIONS:
## "Greetings from The Lyme Cult"

"Great abuses in the world are begotten... by our being taught to be afraid of professing ignorance."
Michel de Montaigne[550]

"I've been in business 42 years doing this, this is easily the toughest disease I've ever seen to diagnose or treat."
Dr. Norton Fishman[551]

Cotton Mather was not totally off the mark when he wrote about the "wonders of the invisible world." [552]He saw in human affliction the work of the hand of evil. His invisible world, we now know, is inhabited by bacteria and viruses that can cause affliction and death. If the hand of evil is redefined as being the results of microbial infection- a new understanding of the past can be formulated. The idea that Lyme disease affected people in the past fits logically with modern scientific research, historic writings, and major scholarly works about the seventeenth century. This is especially true for the witchcraft episode of 1692.

Alan MacFarlane argues in *Witchcraft in Tudor and Stuart England* that witchcraft outbreaks occur historically during periods where societies are evolving from a communal to an individualistic ethic.[553] In northern deciduous forested areas the most important environmental effect of this type of social movement is the tendency for the particular culture involved to spread out spatially-often onto land that is already inhabited by an established *Ixodes* tick population. This may be true of several of the witchcraft outbreaks that occurred in Europe. In Germany and France they occurred during periods of population expansion along the Rhine and Rhone River valleys. In England, the social and legal changes in land ownership and use after the Norman Conquest of 1066 may have had the environmental result of reforested fragments being created as upper class hunting parks. The sudden imposition of upper class ownership onto wild mammals, most notably deer and boar, may have led to population fluctuations that benefited the *Ixodes* life cycle.[554]In New England, Native American culture had evolved a subsistence system that effectively controlled *Ixodes* ticks in habitated areas over many centuries but it did not protect them from European diseases like

## DISGUISED AS THE DEVIL

smallpox, measles, and the plague. The early English settler's activities related to commerce, usually based on an individualistic desire for profit, moved people and their animals out of the modified environments of Plymouth, Boston and Salem Town into forested edge ecotones and sometimes into the forest itself. The afflictions of Lyme disease may be an almost unavoidable "side effect" to any deforestation movement in northern deciduous forested areas. The neurological nature of some of Lyme disease's symptoms fits almost seamlessly with the diagnosis of preternatural affliction at these times.[555]

English settlers in seventeenth-century Massachusetts interacted with the landscape. They perceived the events of their lives through the lens of their own particular culture. Women and children wore their culturally prescribed long skirts that were inadvertent tick collectors. They used brass or iron and brass pins that coincidentally were the same color as the *Ixodes scapularis* tick. If a circular rash developed on their bodies, it was easily perceived as a sign of the scurvy or a preternatural bite mark. Neurological afflictions that challenged the mind for comprehension were perceived of as the strange and cunning work of an invisible demonic realm. Seizures and lameness were part of the etiology of witchcraft affliction. The defensive response to human agony in 1692 caused the Salem Village society to crack along an east/west dividing line. Like a flawed ceramic pot that was made with a hairline crack, once it split it would be difficult to put back together in a new way that made sense.[556]

Everything that happened in life needed to be fit neatly into the way that the world was mentally ordered for these English citizens. Bacteria caused their hops to become beer. It curdled their milk. They lived their lives surrounded by bacteria but were unaware. It would be centuries into the future before the germ theory for disease fit neatly into anyone's mental world order. And we still have much to learn about the workings of this invisible bacterial world.

Lyme disease is caused by a bacterium that has both a complex life cycle and an extremely successful evolutionary history-it parasitizes but usually does not kill its hosts. Accepting an extended history for this disease would challenge some of our most basic perceptions. It may be that we have lived with this bacterium for eons and that it helped shape our social history as much as our social history has responded to it. Women and young children, who were at such a disadvantage in the past because of the style of clothing worn, were likely to suffer from both the acute and the chronic form of Lyme disease in greater numbers than their male

## CONCLUSIONS

counterparts. This Lyme induced 'weakness' may have carried a social price with it. Woman bore the double burden of both witchcraft accusations and more affliction.

The anti-fertility properties of Lyme disease, especially increased spontaneous abortion rates, seem to be embroidered into tales of witchcraft from the past. Other symptoms and facts concerning the dissemination of the Lyme bacteria seem to be subtlety encoded into witchcraft folklore. In witchcraft an otherworldly thing (be it devil or tick) attaches to the body in hidden places like "under the left arm" or "on her right side near her ribs" or in "the more secret parts" and suckles blood.[557] The witch has a mark on her body that is numbed and feels no sensation. The Lyme victim also does not usually feel an attached tick because of the anesthetic quality of tick saliva. A round bull's eye rash sometimes develops. In 1692, some of the afflicted developed round rashes on various parts of their bodies that are almost always described as being round like a bite mark.

It is important to remember that in 1692 the society that inhabited Massachusetts Bay was at war. Salem Village had absorbed a stream of human refugees from Maine. The stress suffered during some of the horrendous events that some of the youngest afflicted girls had experienced must have been nearly unbearable. Stress can have a strong affect on Lyme disease victims. It can reawaken dormant spirochetes in birds. It may have exacerbated the symptoms in victims like Mercy Lewis and Sarah Churchill who had fled from the warfare that was raging to "the Eastward" in Maine.[558]

There are frustratingly unattainable 'facts' that may have made for a high rate of Lyme disease infection in 1692. The forest may have experienced an "acorn year" in 1690. The pigs may have been slaughtered or sent away with the invading force to Quebec. This left abundant acorns to nourish a drought concentrated mouse population explosion, which triggered a subsequent higher than normal rate of Lyme disease. The trickle of witchcraft accusations that had occurred throughout the settlement period turned into a torrent of disease and upheaval. The devastating fires that were set in Maine may have also generated an influx of hungry, thirsty, refugee deer that migrated down into the relatively more tranquil areas of Andover, Ipswich, or Salem Village. Deer are skitterish creatures that flee from fire and the retort of guns.[559]

Lyme disease may have affected the early demographics of both the Plymouth and Massachusetts Bay colonies. By initially sickening and depleting the human population, it caused significant delays in development at Plymouth, Salem, and Boston. The death of almost all the women in Plymouth Colony during the winter of 1620-21 led to an abrupt halt in reproduction and an almost total dependence on emigration (especially of fertile women) for any future growth at all. Reproductive rates were also decreased in the Massachusetts Bay Colony where many women died. By infecting cows, Lyme disease may have been an important triggering factor behind the original expansion of Puritan settlers into the Connecticut River Valley. In the mid 1630's this move was highly debated before the Massachusetts General Court but was allowed. This created an inland expansion of the English culture area in New England that would never be turned back. The town of Lyme, Connecticut, which plays an important role in the modern story of Lyme disease itself was founded during this early expansion.

Unlike other more virulent diseases, the history of Lyme disease has been a subtle trail of affliction that did not begin to be defined in America until the nineteen seventies. In context with other more deadly diseases, it was easily missed. But even with the brutally high mortality rates, for example, that the plague created for centuries, it took until 1894 (540 years after it killed off more than thirty percent of Europe's population) for a possible responsible bacteria, *Yersinia pestis,* to be discovered, and three more years for the flea and rat vectors to be determined. This modern insight into the mechanism of spread for the Black Death was then circumstantially applied to events that had occurred centuries earlier.[560] It took until 2004 for DNA amplification of material from plague pit burials in England and France to verify that some of the dead had indeed been infected with *Yersinia pestis.* Similar research is now being done on the remains of victims from Justinian's plague of 600 A.D.[561]

A historical search of cultural customs and patterns that emerged in post plague Europe has only enhanced our understanding of society's reactions to disease. What our culture learned from the Plague Era's quarantine has been used to control the spread of SARS. The same cats, terriers and ferrets that were put on plague patrol to control rats in early modern Europe remain our twenty first century companions as house pets.[562]

The same type research should be done for Lyme disease because, while not as deadly, it is an infectious factor that has had arguably profound affects on human history. By accompanying Columbus and other Spanish explorers back to Europe after 1492, a virulent strain of *Bb* related sickness may have sparked an up-tick in the great witch-hunts of the

## CONCLUSIONS

sixteenth century. The characteristics of the disease, especially the ECM rash that was interpreted as the devil's or witch's mark, defined the stereotypical European witch. By afflicting people in the Salem Village area, Lyme disease may have caused reactive violence to be inflicted in the name of the government of Massachusetts Bay, which in turn created a cultural response that has never really ended. Our society's reaction to the use of torture to elicit information from prisoners remains strong. The division between church and state that was incorporated into the United States Constitution, as well as the inclusion of personal property rights and protection against self incrimination in the Fifth Amendment in the Bill of Rights are in part a reaction to the events of 1692.[563] And the "Witchcraft Hysteria" is still the subject of modern fascination, research, and scholarship.

As the United States grew, the English cultural area spread out of New England into the Mid-Atlantic States, the upper areas of the Mid-West, and eventually to parts of the Pacific Coast, especially the gold rush areas around San Francisco that were heavily colonized out of New England. All of these areas had high rates of both prevalence and risk for Lyme disease in 2002. Comparing a map of the pre-civil war population expansion out of New England with a modern Lyme disease epidemiological prevalence map shows that there are some interesting overlaps of geographical range. Could some of the *Borrelia burgdorferi* bacteria have been spread by humans and their domesticated animals when they moved to farms in the Midwest or retired to Florida? This underscores a need for further research into the topics of human vector competence and sexual and congenital transfer as possible amplifying factors in Lyme disease's epidemiological transmission models. Far from being a modern, easily curable disease of insignificant effect that readily responds to antibiotics, Lyme disease may hold an important, complex, and continuing role as a variable factor of influence in America's cultural history.

Accepting the fact that Lyme disease has a past may give modern researchers new insights into the present. Meeting the *Healthy People 2010* [564] goal of a forty percent decrease in the prevalence of Lyme disease in the United States will take a Herculean effort. It will require a shift in approach. The many symptoms that Polly Murray and her family presented to Yale's New Haven doctors in the 1970's reflected the complexity of this disease and did not fit into any prevailing medical definition or a diagnostic model. Even today, Lyme disease does not fit into a medical model for making a definitive diagnosis. What Polly

## DISGUISED AS THE DEVIL

Murray faced twenty-five years ago continues today for many people with Lyme disease. The prevailing strategy seems to be: control medical costs by denying the complexity of the disease. Like the witch-hunts of old, the modern sufferers of the chronic form of this disease have been accused of belonging to a mythical Lyme "Cult" that threatens the very stability of the status quo. For those who have been affected by what may be the most devastating form of this affliction, the internet provides a forum, a community and a means of voicing ideas that would otherwise be silenced. There appears to be a growing group of previously healthy people who have suffered for months or years after initial treatment to suggest that there is often more to Lyme disease than "official" diagnostic and treatment guidelines suggest.

If Lyme disease is seen as an old affliction, it might make late or chronic Lyme disease and neurological Lyme disease the focus of more research. Better tests and reporting methods are needed to develop true prevalence rates. The CDC's own formula of one reported symptomatic case for every ten actual occurrences, and then the doubling of it by using Dr. A. Steere's own early finding of one asymptomatic case for every symptomatic case, gives an astounding infection rate for most endemic areas of modern America!

"The Burning Times" of the great witch-hunts of the past have been held up by many members of modern society as emblematic of ignorance and intolerance in the past-societal actions to be avoided, not emulated. Unfortunately, many of the techniques and exclusionary tactics learned during the witch-hunts of the past continue to be used as tools to stifle dissent in the present. The revival of water boarding for accused terrorists and disciplinary reviews designed to bankrupt medical practitioners with astronomical legal fees come quickly to mind.

Like the witchcraft afflictions of old, Lyme disease has become politicized. Over the past decade, two opposing camps supporting two contrasting paradigms have emerged. One camp is represented by the Infectious Diseases Society of America (IDSA), which maintains that Lyme disease is a rare illness that is localized to well-defined areas of the world. According to the IDSA, the disease is 'hard to catch and easy to cure' because the infection is rarely encountered, easily diagnosed in its early stage by means of accurate commercial laboratory tests, and effectively treated with a short course of oral antibiotics over 2-4 weeks. Chronic infection with the Lyme spirochete does not exist.

Like the witch-hunts of old, the politics of this disease has created a full set of victims. For example, in a recent speech Dr. Steere describes

## CONCLUSIONS

people who "believe" in the chronic form of the disease as a "Lyme counterculture that has emerged" and bemoans the fact that we are "drifting in space" because people are leery of Lyme vaccines. He failed to tell his audience that he would profit handsomely from the sale of the vaccines that he holds patents to, and went on to compare himself to Galileo.[565] This may have been an unfortunate selection, because Galileo himself was a victim of an inflexible Inquisition and suffered for questioning rote dogma. He had little to gain other than being found to be right in the judgment of history. He was never on the payroll of a HMO and he died a poor man.

The tactics used today by those who deny the existence of Chronic Lyme more closely mirror those of Galileo's Inquisitors. Doctors are labeled as "quacks" and reported to state medical boards for licensing reviews. The patients who continue to exhibit symptoms, especially those of a neurological nature, are easily labeled as crazy or malingerers. Chronic Lyme disease places people into a form of house arrest-too sick to carry on with a normal life with an affliction that they are told they must deny! Even worse, doctors who treat sick children who are not cured after two weeks of antibiotics are being accused of falsifying an illness in a child in order to justify absences from school, essentially colluding with parents, who are accused of Munchausen's Syndrome by Proxy.*

An opposing paradigm is represented by the International Lyme and Associated Diseases Society (ILADS), which argues that Lyme disease is not rare and, because its spread is facilitated by rodents, deer and birds, it can be found in an unpredictable distribution around the world. It is often accompanied by other co-infections that further complicate the clinical picture. According to ILADS, tick bites often go unnoticed and because commercial laboratory testing for Lyme disease is inaccurate, the disease is often not recognized and may persist in a number of patients. They may require prolonged therapy to eradicate persistent infection with the evasive Lyme spirochete.

It is the Lyme-Literate practitioners who follow the ILADS guidelines that are following in the true footsteps of Galileo. The battle over chronic Lyme disease has taken some unprecedented turns. As of 2007, more than 19,000 scientific articles about tick-borne diseases have been published, and the dichotomy between the basic science studies and clinical research articles is striking.

*as is currently happening in Connecticut, 5/3/2008.

Basic science studies continue to highlight the invasiveness and elusiveness of *Bb,* while clinical research articles adhere to a strict dogma that *Bb* produces a limited infection that is easily eradicated. Patients with persistent symptoms are labeled as having 'post-Lyme syndrome' which is hypothesized to be just another autoimmune response or a form of mental illness. While IDSA followers have embraced the post-Lyme syndrome concept, which limits antibiotics and treatment and is less costly, followers of the ILADS have continued to use antibiotics to treat persistently symptomatic Lyme patients for chronic infection. They cite animal studies that demonstrate persistent infection by this complex organism, as well as numerous clinical reports that document failure of the standard two to four weeks of antibiotic therapy recommended by the IDSA.

At present, the privately developed IDSA guidelines and policies are supported by the publicly funded CDC, to the exclusion of all others. The IDSA functions much like the one party theocracy of seventeenth century Massachusetts. There appears to be no mechanism for handling dissenting opinions other than ostracism and animosity. The academics on the IDSA side of the battle dominate the marketplace by holding positions of influence both inside and outside of government and then use those powerful positions to further the IDSA paradigm. They have tried to control almost every aspect of Lyme disease: research, testing methods, diagnostic standards, treatment standards, insurance reimbursement and even the livelihoods of those doctors who treat outside the IDSA protocols. They even try to control patients by labeling them as an internet "Lyme cult."

Because of their connections at NIH and CDC, most of the public grant money available for Lyme disease research has been focusing heavily on vaccines. Their sphere of influence extends to medical journals, where they are able to impact which studies get published and which studies don't, through a "peer review" process. They have similarly used their powerful connections to encourage and support investigations by state medical boards against doctors who treat chronic Lyme disease. They have used their consulting relationships with insurance companies to deprive patients of access to antibiotic therapy and/or insurance reimbursement, even when that therapy is prescribed by a licensed physician and has been shown to improve the patients' health.

In November 2006, the IDSA published an updated set of diagnostic and treatment guidelines that literally defined chronic Lyme disease out of existence. A link to the list can be found at the CDC Lyme disease website at www.cdc.gov. This has brought unrelenting misery to

## CONCLUSIONS

thousands of patients. It may represent the tipping point for this disease. Questions have been raised about the scientific objectivity of these guidelines. The authors relied on only 405 of the "more than 19,000 scientific studies on tick-borne diseases" when they formulated their conclusions and they largely disregarded any contradictory research. In fact, a significant percentage of the 405 studies that were used were authored by the same people who were writing the IDSA guidelines. Many of these authors have financial ties to insurance companies, labs, and Lyme related patents, royalties, etc. which might have provided motivation for such bias. The IDSA guidelines are so restrictive that the Attorney General of Connecticut initiated an unprecedented investigation into possible antitrust violations in the formulation of the guidelines.

The IDSA guidelines were soon followed by the publication of a similar set of guidelines from the American Academy of Neurology (AAN). Although these new guidelines were presented as independent scientific corroboration of the IDSA guidelines, it turns out that the AAN guidelines were authored by three of the same people.[566] Connecticut Attorney General Blumenthal has now also subpoenaed related documents from the AAN.[567]

On October 4, 2007, the New England Journal of Medicine (NEJM) joined the fray on the side of the IDSA by publishing a research review entitled, *A Critical Appraisal of "Chronic Lyme Disease."* That article reached virtually the same conclusions as the previously released IDSA and AAN guidelines: that Chronic Lyme disease is a misnomer, and the use of prolonged, dangerous, and *expensive* antibiotic treatment for it is not warranted. This was not, however, original Lyme disease research in any way, shape or form. The "study" was a literature review of the previous studies that had already been used to formulate the IDSA guidelines.

The NEJM study's listed authors included eleven of the fourteen members of the panel that wrote the IDSA guidelines. The NEJM also neglected to mention the fact that IDSA guideline author, Dr. Mark Klempner, is also an associate editor at NEJM. This may well have influenced the decision to publish this article. The conflicts of interest that were mentioned included: lecture fees from Merck Pharmaceutical Company, serving as an expert witness in medical-malpractice cases related to Lyme disease, holding patents on diagnostic antigens for Lyme disease, serving as an expert witness related to Lyme disease issues in civil and criminal cases in England, serving as an expert witness in medical-malpractice cases related to Lyme disease, reviewing claims of disability related to Lyme disease for Metropolitan Life Insurance Company, receiving speaker's fees from Merck and Sanofi-Aventis, a research grant from Viramed and fees from Novartis, research grants related to Lyme disease from Immunetics, Bio-Rad, and Biopeptides,education grants from Merck and AstraZeneca for visiting lecturers, being part owner of Diaspex, owning equity in Abbott, serving as an expert witness in a medical-malpractice case, and, being retained in other medical-malpractice cases involving Lyme disease.

  Extracurricular activities of the type listed, most notably those related to the medical insurance industry, give the appearance of conflicts of interest. With the quest for ever greater profits motivating insurers to deny payment wherever possible, it is only natural that they would seek out Lyme related medical opinions from those experts who are most willing to deny the existence of the most expensive form of the disease. Cementing the IDSA author's control over the definition of Lyme disease guarantees that these medical insurance benefactors will *never again have to pay* for any expensive long-term treatments for this disease.

  This makes for a fierce battle over treatment. Some doctors are being hauled before their medical boards for over-prescribing-especially for expensive IV antibiotics. People are begging their vets for the treatment options given to their dogs because they can not get treatment for themselves. ILADS believes that twenty years of scientific studies back up their standards but that this research is being ignored.

On May 1, 2008, Attorney General Richard Blumenthal announced that his antitrust investigation had uncovered serious flaws in the IDSA process for writing its 2006 Lyme disease guidelines and that the IDSA had agreed to reassess them with the assistance of an outside arbiter. Blumenthal wrote:

  "This agreement vindicates my investigation- finding undisclosed financial interests and forcing a reassessment of IDSA guidelines…My office uncovered undisclosed financial interests held by several of the most powerful IDSA panelists. The IDSA's guideline panel improperly ignored or minimized consideration of alternative medical opinion and evidence regarding chronic Lyme disease, potentially raising serious questions about whether the recommendations reflected all relevant science. The IDSA's Lyme guideline process lacked important procedural safeguards requiring complete reevaluation of the 2006 Lyme disease guidelines- in effect a comprehensive reassessment through a new panel. The new panel will accept and analyze all evidence, including divergent opinion. An independent neutral ombudsman – an expert in medical

## CONCLUSIONS

ethics and conflicts of interest, selected by both the IDSA and my office – will assess the new panel for conflicts of interests and ensure its integrity."

Blumenthal's findings included the following:

The IDSA failed to conduct a conflicts of interest review for any of the panelists prior to their appointment to the 2006 Lyme disease guideline panel; Subsequent disclosures demonstrate that several of the 2006 Lyme disease panelists had conflicts of interest;

The IDSA failed to follow its own procedures for appointing the 2006 panel chairman and members, enabling the chairman, who held a bias regarding the existence of chronic Lyme, to handpick a likeminded panel without scrutiny by or formal approval of the IDSA's oversight committee;

The IDSA's 2000 and 2006 Lyme disease panels refused to accept or meaningfully consider information regarding the existence of chronic Lyme disease, once removing a panelist from the 2000 panel who dissented from the group's position on chronic Lyme disease to achieve "consensus"[568]

The IDSA blocked appointment of scientists and physicians with divergent views on chronic Lyme who sought to serve on the 2006 guidelines panel by informing them that the panel was fully staffed, even though it was later expanded;

The IDSA portrayed the American Academy of Neurology's Lyme disease guidelines as corroborating its own when it knew that the two panels shared several authors, including the chairmen of both groups, and were working on guidelines at the same time. In allowing its panelists to serve on both groups at the same time, IDSA violated its own conflicts of interest policy.

Blumenthal added, "The IDSA's 2006 Lyme disease guideline panel undercut its credibility by allowing individuals with financial interests to exclude divergent medical evidence and opinion. In today's healthcare system, clinical practice guidelines have tremendous influence on the marketing of medical services and products, insurance reimbursements and treatment decisions. As a result, medical societies that publish such guidelines have a legal and moral duty to use exacting safeguards and scientific standards."[569] The IDSA has admitted to no wrongdoing.

One of Connecticut Attorney General Blumenthal's concerns is for something that is already happening: medical insurance companies use the IDSA guidelines to refuse to pay for prescribed medications beyond four weeks. Thousands, with thousands more to come; have been caught up in this medical melee.[570] Eventually it will become such a problem that we may end up with a whole new class of destitute homeless people on the streets in America: the chronically ill who are deprived of medical care!

The role of antibiotics as a cure for this disease may justifiably need to be carefully scrutinized. Nevertheless, it needs to be part of a holistic re-examination. Why are Lyme patients being singled out when acne patients are not, and even worse, when the majority of antibiotics sold in the United States are pumped directly into the food stream when they are routinely fed to factory farmed animals? Dogs and cats are regularly and readily prescribed antibiotics when they come down with this disease.

The imperative of early treatment is acknowledged by everyone. But the fact that the current diagnostic blood tests often do not even turn positive until long after "early" treatment is possible is a problem. Inadequate initial doses of antibiotics may in reality be doing little more than converting the spirochetes into a cystic state where they can persist, causing a lifetime of future affliction for some patients. [571] Saving the insurance companies a few dollars at the onset of infection by limiting treatment may result in higher societal costs in the future: by creating drug resistant forms of the bacteria, lost wages, medical care, and human suffering.[572] Doctors who aggressively treat and actually report cases to the CDC should not be subjected to harassment or censured by Health Maintenance Organizations.

The earlier hope for a safe vaccine seems to be unrealistic, especially as our understanding of the extremely complex and relapsing life cycle for the *Borrelia burgdorferi* bacteria comes more clearly into focus. The role of the immune system in this process is not fully understood. The chronic but usually controllable nature of the infection needs to be accepted and treated. Immediate research strategies need to

## CONCLUSIONS

shift away from literature reviews designed to undermine the lives of chronic Lyme patients and the dogged pursuit of lucrative vaccines, and refocus on preventing infection in the first place. It appears that this will be more difficult than it seems. Moreover, sometimes even this research seems to be misdirected.

One of the latest NIH funded studies to be released suggests that we may be able to decrease the number of infected nymphal ticks in the forests of New England by twenty-seven percent by vaccinating the entire wild white footed mouse population. The logistics of this new plan are difficult to fathom. There must be millions of mice that would need to be trapped, vaccinated, and released. If this new mouse vaccine is anything like the old ones, it will require annual booster shots, which would require more trapping and probably some form of tagging system in order to figure out which mouse had gotten its shot. It would certainly be less bother to just trap and kill these rodents. But then the pharmaceutical companies would not be able to supply and profit from millions of doses of mouse vaccine. The fact that this scheme is being presented as a viable approach to the Lyme disease problem is absurd.[573] It just shows the extent that some will go to turn a buck. The absurdity of America's antibiotic usage policies is revealed in another study that proposes feeding antibiotic laced bait to rodents to reduce Lyme disease. In this scheme, the forest floor would be littered with the doxycycline that is so precious that it is being withheld from human patients for long term treatment. Perhaps some Lyme patients will resort to scouring the forest floors for mouse bait in order to obtain antibiotics! [574]

The acceptance of a long history for Lyme disease may also shed some much needed light into the area of mental illness. One blood sample survey done in a modern mental institution found that an astounding forty percent of the patients tested positive for Lyme infection.[575] With the invention of brain image scanning, we can only now actually begin to *see* the damage that these spirochetes do to the brain itself. This gives additional credence for the infective nature of some dysfunctions of the brain.

What we truly need to do is to decrease the tick population in endemic areas and do it in a way that does not scourge our bodies and the earth with pesticides or be formulated simply to improve the bottom line of the drug companies. Looking at old and sustainable ideas like clustered housing, regular leaf burning, pannage, and acorn removal could be coupled with experiments that find ways to use acorn derivatives as bio-fuel or building materials. Removing acorns in areas where humans live would help limit the population of mice and deer by removing a primary food source. While it is controversial and not likely to happen, the return of the native predator, the wolf, would be an excellent mouse and deer population control measure. We may also need to be surrounded by more domesticated animals that graze once again. As one study has suggested, a return to traditional dispersed non-intensive agricultural herding practices would also help reduce the risk for Lyme disease. Looking backward to understand the ways that earlier inhabitants of tick risky areas interacted with their environment, lived their lives, and coped with disease may help to plot the course forward.[576]

During the writing of this book a particularly beloved songwriter who lived in a nearby town committed suicide. He had suffered from chronic Lyme disease for years. His death was a devastating blow to his friends and family and caused the importance of this disease to hit home. The fact that some people who are very, very, very sick have been caught up in an acrimonious form of paradigm wars and are losing all hope is not acceptable. The fact that patients have to beg for treatment is not acceptable. As one member of a Lyme disease support group said to me, on learning of the topic of this book, "at least a witch died fast when they were burned. My death will be slow, drawn out, and filled with daily pain and suffering. I think that I would rather be hung." Another member of a support group cannot understand why the physicians and veterinarians in her geographic area give such different medical advice on long term treatment, "Oh," she mused, "to be a dog! Then I would be adequately treated!" And even if they can find a Lyme literate physician; patients are sometimes denied medications under their medical insurance even if it has been prescribed by that licensed physician. These physicians sometimes suffer under constant attack. On top of all this, almost all pain control is approached with the assumption of presumed addiction.

Modern medicine in the United States is a for-profit world. The same politicians who have pronounced that socialized medicine is a failure and that medical decisions should be between a citizen and his physician have willingly handed over that power to the Medical Insurance Industry over the past several decades! Chronic Lyme disease does not fit into this model very well because it requires both acknowledgement and then payment for protracted treatments. Health care is no longer a common right in America. I also find it troubling when a group of medical professionals who are increasingly tied financially to the insurance companies are allowed to formulate guidelines that are

## CONCLUSIONS

dismissive of what may be the most devastating form of any disease. Repeating the mantra that chronic Lyme disease does not exist over and over again does not make the information true-or updated-or even research. There are too many who have suffered for too long to be ignored!

It is also troubling that Lyme disease researchers have now latched on to homeland security bio-weapon research funding. IDSA members sit on the Epidemic Intelligence Service and lead both a multi million-dollar bio- warfare research center at the University of California Irvine and the 1.6 billion dollar bio-warfare mega complex currently being built, under protest, at Boston University. It makes one wonder if we have unwittingly been recruited as participants in the study of long-term effects of untreated Lyme disease or, worse, what will happen if accidents occur. Will they be acknowledged or will the infected be told that they are not sick and be left to fend for themselves on the streets of America.

We need medical and political leaders who are willing to say that there is a better way and who stand up for what they say. Lyme disease, like autism and other particularly thorny afflictions, is a test for both our medical and our political system. The cliché that the future of our democracy is at risk may apply now more than ever. I worry about Lyme Disease bills that get passed in Congress without any funding but I also worry about any future bills that may get passed and are finally funded only to have the monies go into the pockets of the very same researchers who want to do things like vaccinate forests full of wild mice or lit reviews of their own work to prove once again that people with chronic Lyme are crazy! We need to focus on human needs first and set better priorities. We need environmental remediation. **We need an accurate early diagnostic test** and **we need compassionate late stage care.** We need to develop strategies for those who continue to suffer from the effects of this disease long after the medical establishment has said that they should be cured. In addition, everyone needs to have a place at the table when the discussion begins and the future is planned. We need to be able to form a consensus of opinion within policy mechanisms that allow for, and even encourage, dissent. The fact that neurological Lyme affects many with a form of mental illness has muddied the picture but adds special urgency to this matter. Since serious mental illness costs Americans at least $193 billion a year[577] in lost earnings alone, if any of that suffering could be alleviated it would be well worth it in both cost savings and improved lives.

DISGUISED AS THE DEVIL

We need to work together to get it right. We need to value all scientific research, not just that which supports one side of the debate. Moreover, if we do, we can create an informed future where the medical research system is centered on need, knowledge, and integrity, not based solely on recalcitrance or even worse, profitability. Medical care should be determined as a holistic and collaborative effort between patient and practitioner and public policy should support this precious relationship. Moreover, we need to stop blaming the victims. No one ever asked for the devastation that the chronic form of this affliction can bring, any more than anyone ever asked in the past to be labeled as a witch or to suffer from witchcraft afflictions. Our disease is not an affront to modern medical practice: it is a challenging puzzle that needs a better solution. **WE CAN AND MUST DO BETTER** in the future. If the needs of the afflicted are met and the cycle of infection is broken, the spread of Lyme disease will be contained. If we get it right and are ever vigilant, we can ensure that this all too real 'affliction with a past' will begin to fly away once and for all, like the witches of old.

**CONCLUSIONS**

# CONCLUSIONS

550 William Carew Hazlitt, Ed., *The Essays of Michel de Montaigne* (1877 Edition) is available online as an e-book at Project Gutenberg.
551 Interview, WABC 13, *The Lyme Controversy* - Part 3 November 08, 2007- Lynchburg, VA.
552 Cotton Mather's book is given this title, full text is in Burr *Narratives of the New England Witchcraft Cases* (2002) that has been used throughout this work.
553 Alan MacFarlane, *Witchcraft in Tudor and Stuart England* (New York: Harper and Row, 1970), 158-164, 192-206.
554 The link of affliction with the Norman conquest of England is noted in Christopher Lee, *1603* (New York: St. Martin's Press, 2003), 120-121. The Normans imposed a series of new property laws, reforested many areas by establishing hunting parks and made wild mammals like deer into the exclusive property of the nobility and the king.
555 Ipswich, which is directly north of Salem Village had similar problems earlier. By 1658 the residents complained that "for some time suffered losses in their estates, and some affliction in their bodies also...which doth not arise from any natural cause" in Boyer and Nissenbaum, eds., *Salem-Village Witchcraft: A Documentary Record of Local Conflict in Colonial New England* (Boston: Northeastern University Press, 1993), 429.
556 The most in depth study of the social aspects of the 1692 event is Paul Boyer and Stephen Nissenbaum's *Salem Possessed: the Social Origins of Witchcraft* (Cambridge, Massachusetts: Harvard University Press, 1974).
557 Dgar Peel and Pat Southern, *The Trials of the Lancashire Witches: A Study of Seventeenth Century Witchcraft* (New York: Taplinger, 1969), 115.
558 Carol Karlsen, *The Devil in the Shape of a Woman* (New York: Vintage, 1987), 227-228. See also Mary Beth Norton's In the Devil's Snare for a lengthy discussion of the warfare connection.
559 See "The Deer with a Collar" chapter in Virginia DeJohn Anderson, *Creatures of Empire* (New York: Oxford University Press, 2004).
560 B. Harvey, ed., *Living and Dying in England* 1100-1540 (Oxford, U.K.: Clarendon Press, 1993), 236-38.
561 Thomas H. Maugh II, "Scientists Use DNA in Search for Answers to Sixth Century Plague," *Los Angeles Times* (May 6, 2002) available at UCLA School of Public Health website www.ph.ucla.edu.
562 *Emerging Infectious Disease*, cover, October 2007
563 In the convoluted legal format of the Salem Witch Trials, a confession of guilt often brought a reprieve from capital punishment and a not guilty plea was often followed by a death sentence. When Giles Corey refused to plead at all he was crushed to death. This part of American history may have directly influenced the inclusion of the Fifth Amendment in the Bill of Rights. It offers protection against self-incrimination. See also Richard P. Conti, "The Psychology of False Confessions," *The Journal of Credibility Assessment and Witness Psychology* 2 No.1 (1999), 14-36.
564 United States Department of Health and Human Services, *Healthy People* 2010 (Washington, D.C.: Government Printing Offices, 2000) and Gregory A. Poland, "Prevention of Lyme disease," *Mayo Clinic Procedures* 76 (2001), 713-724.
565 Dr. Steere's Commencement Address at Ohio Wesleyan University, May 11, 2008 available online at http://stream.owu.edu.
566 Wormser, Shapiro, and Halperin.
567 See "Guidelines on Trial: AAN Subpoenaed as Part of Investigation into Treatment Parameters for Lyme disease" in Steven Gottschalk's article in the *Connecticut Patriot*, Nov 12, 2007. The guidelines can be found in the October 16, 2007 AAN publication, Neurology Today.
568 Dr. Sam Donta
569 Attorney General Richard Blumenthal , "Attorney General's Investigation Reveals Flawed Lyme Disease Guideline Process, IDSA Agrees To Reassess Guidelines, Install Independent Arbiter," *Press Release*, May 1, 2008.
570 Article by Steven Gottschalk, "Attack of the Chronic Lyme Denialists," *Connecticut Patriot,* 11/12/2007. See also "Expert review of Anti-infective Therapy" by Dr. Raphael Stricker, on the ILIADS website.
571 See E. Hodzic, Feng S, Holden K, Freet KJ, Barthold SW. "Persistence of Borrelia burgdorferi Following Antibiotic Treatment in Mice," *Antimicrob Agents Chemother*. Mar 3, 2008, for a study done on mice. This study found that even after antibiotic treatment some mice remain infected with Bb, especially those who were allowed to obtain a chronic form of the disease. Even infection free Ixodes scapularis ticks who were placed on "treated" mice for a blood meal were infected with the bacteria.
572 See Infection, 24 (1996), 218-26 for a complete description of this process.
573 The entomologist Durland Fish is one of the authors of this study. See the National Institute of Allergy and Infectious Disease News Release (12/04/2004) "Broad- based Vaccination of Wild Mice Could Help Reduce

Lyme Disease Risk in Humans" at www.niaid.nih.gov.
574 See Marc C. Dolan, Nordin S. Zeidner, Elizabeth Gabitzsch, Gabrielle Dietrich, Jeff N. Borchert, Rich M. Poché, and Joseph Piesman, "A Doxycycline Hyclate Rodent Bait Formulation for Prophylaxis and Treatment of Tick-transmitted Borrelia burgdorferi," *Am. J. Trop.* Med. Hyg., 78 (5), 2008, 803-805
575 Denise Lang, *Coping with Lyme disease* (New York: Henry Holt, 2004), 15.
576 D. Richter and F. Matuscka, *Modulatory effect of cattle on risk for Lyme disease*, U.S. Center for Infectious Disease (2006) online at www.lymenet.org. See also *Are sheep effective tick mops?*, The Game Conservatory Trust, Scotland.
577 Ronald C. Kessler, Steven Heeringa, Matthew D. Lakoma, Maria Petukhova, Agnes E. Rupp, Michael Schoenbaum,, Philip S. Wang., and Alan M. Zaslavsky, "Individual and Societal Effects of Mental Disorders on Earnings in the United States: Results From the National Co-morbidity Survey Replication," *The American Journal of Psychiatry*, May 7, 2008 accessed online at www.ajp.psychiatryonline.org.

# APPENDIX A

## Appendix A: Seventeenth Century Clothing

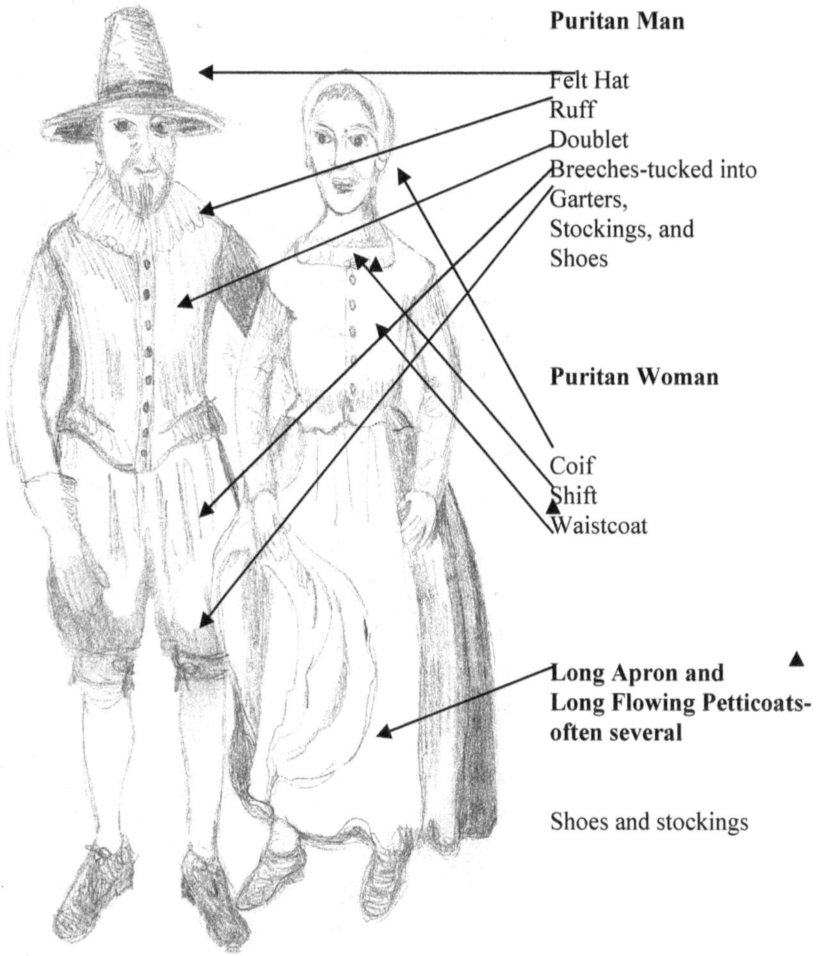

**Puritan Man**

Felt Hat
Ruff
Doublet
Breeches-tucked into
Garters,
Stockings, and
Shoes

**Puritan Woman**

Coif
Shift
Waistcoat

Long Apron and
Long Flowing Petticoats-
often several

Shoes and stockings

The male attire of the seventeenth century closely resembles the modern tick protection recommendations. [see Preventionguy] Women, however, were at a disadvantage in tick risky areas because of the clothing that they wore. The long flowing skirts acted as tick collectors. Women wore long skirts in America until the 1920's.

221

# DISGUISED AS THE DEVIL

# APPENDIX B

## ENGLISH CULTURAL PRACTICES

Cotton Mather revealed his cultural mindset when he wrote about the exceptionality of New England revealing why he thought that the area had been singled out for witchcraft affliction:

*"The New Englanders are a People of God settled in those, which were once the Devil's Territories; ... The Devil thus irritated, immediately try'd all sorts of methods to overturn this poor plantation: .... I believe, that never were more satanical devices used for the unsettling of any people under the sun, than what have been employ'd for the extirpation of the vine which God has here planted, ... But, all those attempts of hell, have hitherto been abortive, .... Wherefore the Devil is now making one attempt more upon us; ... We have been advised by some credible Christians yet alive, that a malefactor, accused of witchcraft as well as murder, and executed in this place more than forty years ago, did then give notice of, an horrible plot against the country by witchcraft, ... which if it were not seasonably discovered, would probably blow up, and pull down all the churches in the country. And we have now with horror seen the discovery of such witchcraft!"*

Mather's mindset existed as part of a full set of English cultural practices that were occurring on a day to day basis within the Massachusetts Bay Colony. They had cultural, social, legal and environmental consequences.

The passage by Cotton Mather is from the original version of *The Wonders of the Invisible World*, as quoted in John M. Taylor, *The Witchcraft Delusion In Colonial Connecticut 1647-1697* (1908) electronic edition available online at www.fullbooks.org.

# DISGUISED AS THE DEVIL

| Cultural Practice or Concepts | Result |
|---|---|
| Normal Diet is Grain based=High Carbohydrate | Switching to High Protein Diet creates fluctuation in blood insulin levels- called starving in Jamestown and Plymouth |
| Use of Matchlock, Not Deer Hunters, prefer beef | High Deer Population= High Tick Population |
| Scurvy was brought to England by Normans. Symptoms include fatigue, joint pain, spots on skin, Appear on long sea journeys and in spring and summer. [Spirochetal illnesses influence Vitamin C depletion] | Both Lyme disease and Vitamin C deficiency called by same name, may be related |
| Pins are made from un-tempered steel, brass or iron with wound brass head that is hammer into a round dot | A pin is the same color and shape as an attached engorged *Ixodes scapularis* tick |

# APPENDIX B

| | |
|---|---|
| Use of straw, eelgrass, corn husks and leaves to stuff mattress ticks, problems with Bedbugs | Potential for ticks to be brought into home in bedding materials |
| Pulling Heavy outer skirt over head and shoulders for warmth or protection from rain | Transfers ticks to the head/neck area which creates higher risk for neurological Lyme symptoms. More likely to occur on chilly fall and spring days |
| Pannage | Prevents acorn poisoning in Cows and horses, deer, mice and birds not attracted by acorns to areas near human habitation-good tick control method |
| Pasturing cattle "in the woods" | Requires frequent human and domesticated animal interaction with edge ecotone |
| Burning leaves and undergrowth in winter and early spring | Excellent tick control |

## DISGUISED AS THE DEVIL

| | |
|---|---|
| Fragmented forest, retention of Woodlots and windrows, planting → | Can be conducive to ticks, deer |
| Orchards | deer attracted to apples |
| Wood collection for cooking and Heating | Humans in tick habitat |
| Dogs and Cats | Ticks brought closer to Humans in fur, dogs kill Wolves, may be vector for *Bb* |
| Military Service causes disability of rheumatism | Specifically links intensive interactions with forest edge areas and lingering disease |
| Transportation by water | Riparian corridors harbor ticks, especially in droughts |
| Women and children wear Long Skirts that brush into vegetation | Drag collecting of ticks |

# APPENDIX B

Wool used for clothing, cloth ⟶ Sheep attract ticks, fleece
Is dyed various colors may be infested with
Ticks when handled by
Humans, dyestuffs may
Be natural materials from
Forest or ecotone areas

Autumn Slaughter/Hunting Brings humans into close
contact with animals at
time of high level of
questing Adult ticks

Observing skin for preternatural Makes skin manifestations
Marks of rash, often described as
Like a round bite mark,
And sucked on area (witch
Teat) part of concept of
witchcraft. Creates a long
list of marks observed on
skin of accused witches
and some of the afflicted
part of the historic record.
May be ECM's/Lyme
Disease.

Sensations of heat or pricking Logically links
Of skin are caused by preternatural neurological symptoms
Forces to Hell and pins

Under the evil hand, behagged, Gives societal validity
And bewitchment are medical to afflictions while
Diagnoses freeing the practitioner
from responsibility for
a cure, makes
neurological symptoms
a component of witch
affliction.

## DISGUISED AS THE DEVIL

Hallucinations are real but ⟶ Allows
"otherworldly" events   spectral evidence
     to be admitted in court

Witchcraft is a capital crime   Facilitates and validates
Punishable by death by hanging   prosecution and execution
     of suspected witches
     for causing physical
     afflictions

# APPENDIX C

## Traditional Anti-Scorbutics: Vitamin C content in foods mentioned in the 17th century English diet

**Ascorbic acid**
**(mg/100g or /100 ml for liquids)**

| | |
|---|---|
| cloud berries | 80 |
| cranberries | 5-10 |
| gooseberries, fresh | 60-65 |
| apple cider | 4-5 |
| scurvy grass (Cochlearia officinalis, leaves and buds) | 200 |
| spruce pine needles | 65-200 |
| spruce (leaves and young shoots) | 30-270 |
| freshly sprouted barley seed | 30-100 |
| malt, dried & powdered | 10 |
| wild leek | 80 |
| onion | 10-30 |
| wild onion grass (*Allium vineale*) | 130 |
| beans, peas | 10-30 |
| lemon | 53 |
| lemon grass | 0 |
| lime | 27 |
| garlic | 12 |
| parsley | 170 |
| liver, kidney | 10-40 |
| rose hips | 1000 |
| shepard's purse *Capsella bursa-pastoris* | 91 |
| strawberry | 86.18 mg per cup |
| fish flesh (cod, char) | 0.5-2 |
| clams | 8.84 mg per clam |
| oyster | 1.78 mg per oyster |

sassafras oil has been found to be carcinogenic and is banned for sale in the USA, although the leaves are still made into the creole gumbo seasoning called file' powder.

## Antimicrobial Activity of Traditional Spices

<u>100 % antimicrobial activity</u>

Garlic, Onion, Allspice

Oregano

<u>75% or more antimicrobial activity</u>

Thyme, Cinnamon, Tarragon

Cumin, Cloves, Lemon grass

Bay leaf, Capsicums, Rosemary

Marjoram, Mustard, Caraway

Mint, Sage, Fennel, Coriander

Dill, Nutmeg

<u>Least effective anti-microbial spices</u>

Basil, Parsley, Cardamom

Pepper (white/black)*

Ginger, Anise seed, Celery seed

Lemon/lime*

* Pepper and lemon act as synergists. The citric acid in lemons has a low PH which disrupts bacterial cell membranes. Pepper has a substance in it that inhibits botulism and is a bioavailability enhancer for other cells.

Source: "Antimicrobial Functions of Spices: Why Some Like It Hot," Jennifer Billing and Paul W. Sherman, *The Quarterly Review of Biology*, Vol. 73, No.1, March 1998, see also "Darwinian Gastronomy: Why we use spices: Spices taste good because they are good for us," *BioScience* Vol. 49 No.6, June, 1999, 453-464.

# APPENDIX D

## Temperature, Moisture and Lyme disease incidence:

Between 1640 and 1700, witches were accused in the English colonies of Virginia, Maryland, North Carolina, New York, Rhode Island, Pennsylvania, New Hampshire, Connecticut and the Maine section of Massachusetts in addition to Massachusetts itself. Of those accused, the majority were female. Witch related affliction might be linked to the weather in an intricate way. Droughts concentrate ticks in riparian areas. When humans also inhabit these same areas, it may cause an increase in Lyme disease. In other areas, humans may benefit from the dry conditions by having fewer ticks and less disease.

### THE WITCH DROUGHT

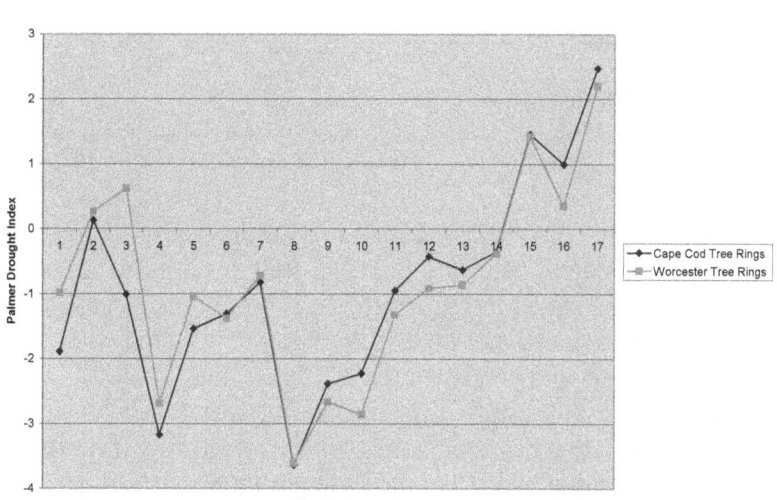

Beginning in the summer of 1681 and for every year until 1692, rainfall in Massachusetts was below average. 1685 was the driest year. This prolonged 'witch drought' must have had devastating effects on agriculture, driven wildlife towards riparian areas, created a risk for forest fires in an era when fire was used as a weapon- and may be one of many underlying causes for King William's War between the English Settlers and Native American populations in the areas to the northeast and west of

## DISGUISED AS THE DEVIL

Salem Village -a classic case of conflict between resident groups over scarce resources in a time of environmental stress.

This period also stands out when placed into a longer timeline. Both the Worcester and Cape Cod areas show a period of drought in the late sixteenth century immediately preceding the English settlement of New England [which affected Native American populations, possibly making them more vulnerable to a variety of diseases] and the more prolonged Witch Drought of the 1680-1692 time period.

**Tree-ring reconstructed Moisture Levels- Worchester, Mass.**

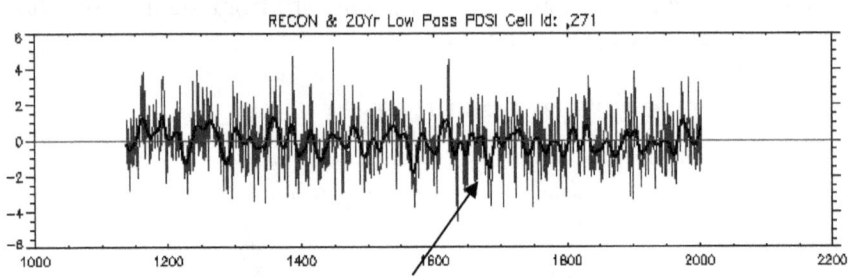

**The Witch Drought**

Graph from NOAA: World Data Center for Paleoclimatology at www.ncdc.noaa.gov/paleo/treering.html

Modern Lyme disease incidence rates for the years 1993-2001 show considerable state-by-state variation. Moisture level influence is felt two years later in some states and sometimes it is not. Lyme disease rates may tend to decrease somewhat with less rainfall, but ticks may then be concentrated near rivers and ponds leading to higher rates in these specific riparian areas. More study is needed in this area.

The research that has been done yields a mixed bag of results. When rainfall as measured by the Palmer Drought Index [PHDI] is compared to Lyme disease incidence rates for several Eastern states, results have been found to be are statistically significant only for three states:

NewYork: $r2 =.74$, $p=0.003$,
Conn.: $r2 =.45$, $p=0.046$, and,
Mass.: $r2 =.49$, $p=0.036$, which means that with a 95% confidence level, 49 percent of the Lyme disease incidence level in Massachusetts in any year is associated with the moisture level that was measured in June of two years prior. The relationship was not significant for three other states that were studied: Rhode Island, Maryland, and Pennsylvania. It is

# APPENDIX D

interesting to note the averaging trend line but also the fact that in modern Massachusetts, a year that had the driest June two years prior also had one of the highest levels of Lyme disease with 800 reported cases.

**MODERN MASSACHUSSETTS ANNUAL CASES LYME DISEASE AT 2 YEAR PRIOR PHDI MOISTURE LEVELS**
# cases

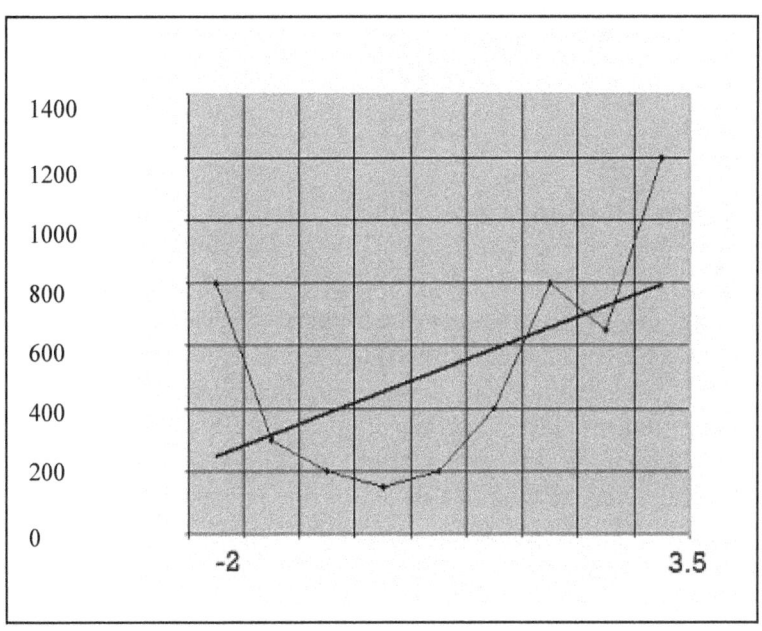

**DRY**                                                           **WET**

Drought index for June in year $t-2$
Actual Number of Reported Cases with Averaging Trendline shows reported cases at -2 equal to reported cases at +2.5.

From information in Subak, Susan (2003) "Effects of Climate Variability in Lyme disease Incidence in the Northeastern United States," *American Journal of Epidemiology* Vol.157, Number 6, 531-538.

 DISGUISED AS THE DEVIL

In addition to being dry, it was also colder than usual between 1680 and 1700, possibly because of volcanic activity.

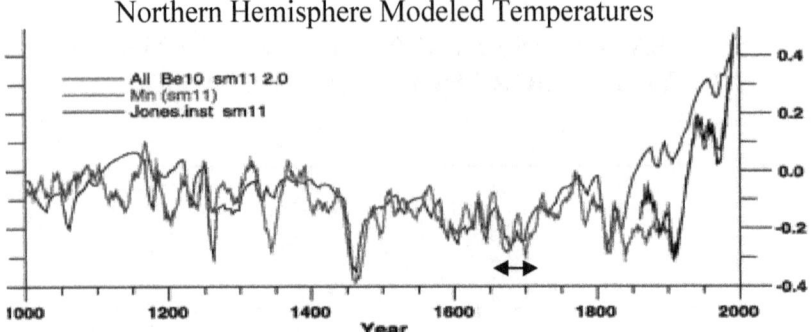

The 1680's began on a worldwide basis with a boom-the eruption of the Tongoko Volcano in Indonesia. This had the effect of cooling temperatures which rivaled only the 1250's and 1450's and would not be felt again until after the 1815 eruption of the Tambora Volcano. A pick up in volcanic eruptions also darkened the skies over Salem in the 1680's and early 90's. 1690 was an unusually active year, there were fourteen volcanoes that had confirmed eruptions during this year, including an eruption in Russia with a VEI of 4, the sunlight blocking effects of this activity would be felt over the next two to three years. Volcanic Explosivity Index (VEI) refers to the amount of particulate matter thrown into the atmosphere. The higher the number, the more dust is thrown into the atmosphere and the more sunlight is blocked.

All information on volcanic activity from www.volcano.si.edu/gvp/usgs.

APPENDIX E

# Modern European Ixodes Tick Infection Rates

Keyed to map on page 88. Lists by region the average percentage of all Infected Nymphs and Adults, with the number of records available.

| Region | Average[%] infected Nymph and Adult *Ixodes* ticks | # Records |
|---|---|---|
| A | 15.2 | 3 |
| B | 5.1 | 5 |
| C | 9.55 | 14 |
| D | 13.15 | 11 |
| E | 8.9 | 6 |
| F | 21.05 | 10 |
| G | 21.25 | 20 |
| H | 21.9 | 62 |
| J | 12.95 | 8 |
| K | 11.4 | 33 |
| L | 8.95 | 7 |
| M | 18.4 | 17 |
| N | 12.4 | 28 |
| O | 6.95 | 5 |
| P | 5.9 | 2 |
| R | 24.25 | 3 |
| S | 25.2 | 5 |
| T | 15.8 | 2 |

## DISGUISED AS THE DEVIL

**Categories for Historic European Witchcraft Accusation Levels: Ratio of executions per capita population set to the year 1600.**

**LEVEL 1**: one in 1-1000 ratio= severest level of accusations. Includes Liechtenstein, Denmark, Belgium/Luxemburg, valley areas of Switzerland, areas now in Poland. In some areas that now make up Germany, there was an average of one witchcraft accusation for every 640 people living there.

**LEVEL 2**: one in 1001-3000. Includes Czech Republic, Slovakia, parts of Scandinavia

**LEVEL 3**: one in 3001-6000. Includes England and Scotland, valleys in Austria, Northern Italy, Finland, and Hungary.

**LEVEL 4**: one in 6000-10,000. Includes The Netherlands.

**LEVEL 5**: one in over 10,000=the lowest level of accusations, areas where witchcraft accusations were practically non-existent (includes Spain, Portugal, Turkey, and Ireland) W. Behringer, *Witches and Witch-hunts*, (Malden, Mass: Polity, 2005).

The association between the level of *borrelia* infection in modern Ixodes tick populations and severity of historic witchcraft accusations is statistically significant. A statistical analysis was made using SPSS 15.0. It found $R^2 =.192$, $P =.047$. It can be said with a 95 percent confidence level that about 19 percent of the chance of being executed as a witch in any area of historic Europe is associated with the severity of infection rates for nymphal and adult *Ixodes* ticks in the same area in modern Europe.

## APPENDIX F
# GROUNDS FOR EXAMINATION OF A WITCH

This list was used for witch trials in Connecticut. It was not dated but the original is written in the handwriting of William Jones, a Deputy Governor of Connecticut from 1692-1698 and a member of the court at some of the trials. It gives a good summary of English witch ideas. I have left it in period dialect.

"1. **Notorious defamacon** by ye common report of the people a ground of suspicion.

"2. Second ground for strict examinacon is if a fellow witch gave testimony on his examinacon or death yt such a pson is a witch, but this is not sufficient for conviccon or condemnacon.

"3. If after **cursing**, there follow death or at least mischiefe to ye party.

"4. If after **quarrelling or threatening** a prsent mischiefe doth follow for ptye's devilishly disposed after **cursing** doe use threatnings, & yt alsoe is a grt prsumcon agt y.

"5. If ye pty suspected be ye son or daughter, the serv't or familiar friend, neer neighbors or old companion of a knowne or convicted witch this alsoe is a prsumcon, for witchcraft is an art yt may be larned & covayd from man to man & oft it falleth out yt a witch dying leaveth som of ye aforesd heires of her witchcraft.

"6. If ye pty suspected have ye **devills mark** for t'is thought wn ye devill maketh his covent with y he alwayess leaves his mark behind him to know y for his owne yt is, if noe evident reason in can be given for such mark.

"7. Lastly if ye pty examined be unconstant & contrary to himselfe in his answers.

"Thus much for examinacon wch usually is by Q. & some tymes by torture upon strong & grt presumcon.

"For conviccon it must be grounded on just and sufficient proofes. The proofes for conviccon of 2 sorts, 1, Some be less sufficient, some more sufficient.

"Less sufficient used in formr ages by red hot iron and scalding water. ye pty to put in his hand in one or take up ye othr, if not hurt ye pty cleered, if hurt convicted for a witch, but this was utterly condemned. In som countryes anothr proofe justified by some of ye learned by casting ye pty bound into water, if she sanck counted inocent, if she sunk not yn guilty, but all those tryalls the author counts supstitious and unwarrantable and worse. Although casting into ye water is by some

# DISGUISED AS THE DEVIL

justified for ye witch having made a ct wth ye devill she hath renounced her baptm & hence ye antipathy between her & water, but this he makes nothing off. Anothr insufficient testimoy of a witch is ye testimony of a wizard, who prtends to show ye face of ye witch to ye party afflicted in a glass, but this he counts diabolicall & dangerous, ye devill may reprsent a pson inocent. Nay if after curses & threats mischiefe follow or if a sick pson like to dy take it on his death such a one has bewitched him, there are strong grounds of suspicon for strict examinacon but not sufficient for conviccon.

"But ye truer proofes sufficient for conviccon are ye voluntary confession of ye pty suspected adjudged sufficient proofe by both divines & lawyers. Or 2 the testimony of 2 witnesses of good and honest report avouching things in theire knowledge before ye magistrat 1 wither yt ye party accused hath made a league wth ye devill or 2d or hath ben some knowne practices of witchcraft. Argumts to prove either must be as 1 if they can pve ye pty hath invocated ye devill for his help this pt of yt ye devill binds withes to.
"Or 2 if ye pty hath entertained a familiar spt in any forme mouse cat or othr visible creature.
"Or 3 if they affirm upon oath ye pty hath done any accon or work wch inferreth a ct wth ye devill, as to shew ye face of a man in a glass, or used inchantmts or such feates, divineing of things to come, raising tempests, or causing ye forme of a dead man to appeare or ye like it sufficiently pves a witch."

But altho those are difficult things to prove yet yr are wayes to come to ye knowledg of y, for tis usuall wth Satan to pmise anything till ye league be ratified, & then he nothing ye discovery of y, for wtever witches intend the devill intends nothing but theire utter confusion, therefore in ye just judgmt of God it soe oft falls out yt some witches shall by confession discour ys, or by true testimonies be convicted.

"And ye reasons why ye devill would discover y is 1 his malice towards all men 2 his insatiable desire to have ye witches not sure enough of y till yn.
"And ye authors warne jurors, &c not to condemne suspected psons on bare prsumtions wthout good & sufficient proofes.
"But if convicted of yt horrid crime to be put to death, for God hath said thou shalt not suffer a witch to live."[578]

---

[578] From John M. Taylor, *The Witchcraft Delusion In Colonial Connecticut 1647-1697* (1908) electronic edition available online at www.fullbooks.org.

# APPENDIX G

# AMERICAN TIMELINE/ AFFLICTED/ GEOGRAPHICAL AREA[579]

[Where symptoms or name of afflicted is unknown, witch, year and geographical area of accusation are listed, if town name only appears it is in Massachusetts]

DATE           Symptoms/Problems           Geographic Area

### 3000-5000 years ago
Late Archaic Period    Erosive lesions in skeletal remains       Alabama
(Rothschild, et al.)

### 300-500 AD
Tchefuncte tribe       Erosive lesions in skeletal remains       Louisiana
(B.A. Lewis)

### Pre-Contact
Algonquin tribes       Rheumatism from Deer              Eastern Woodlands
Cherokee               Erosive lesions in skeletal remains    Tennessee, Kentucky
(Oral History:
Krech, Rothschild)

### English Settlement
### 1607
Jamestown settlers     Irritability, fevers, lethargy, laziness,     Virginia
(John Smith, General            swellings, death
Historie of Virginia, 1624).

### 1620

Plymouth settlers      Starvation, scurvy, fevers, joint pain,    Massachusetts
(Bradford, Winslow)         lethargy, sore throat, death

### 1622-26
Joan Wright            midwife, witchcraft accusations     Jamestown, Virginia
(Davis)

### 1628-30
Massachusetts Bay     Scurvy, fevers, lameness, lethargy,   Massachusetts settlers
(Winthrop, Higginson, Wood)          death

### 1630's
Massachusetts and     Bad grass, spontaneous abortion        Connecticut
Connecticut settlers       in cattle, scurvy
(Winthrop)

## DISGUISED AS THE DEVIL

**1638**
Jane Hawkins   midwife, accused witch, Antinomian heresy   Boston
(Schutte)

**1640**
Ann Hutchinson,   Monstrous birth, Antinomian heresy   Boston and
Collins, Hales    Antinomian heretic                   Aquiday (Portsmouth)
                                                       Rhode Island
                                                       Boston

Mary Dyer ?   Monstrous birth, "relationship with devil"   Romney Marsh
              Antinomian heretic
(Schutte)

**1641**
Witchcraft accusations   Virginia
(Davis)

**1645**
Witchcraft accusations   several "were disturbed"   Springfield
(Drake)

**1647**
Alice (Alse) Young   accused witch   Windsor, CT
(Taylor)

**1648**
Margaret and Thomas Jones   cunning folk healers   Charlestown
"such a malignant touche, as many persons (men
Woemen & children) whom she stroked or touched ...
were taken with deafnesse, or vomitinge or other
violent paynes or sickness. She had a witches teat
"in her secrett partes, as freshe as if it had been newly sucked."
(Winthrop, June 1648)

Mary Johnson   witchcraft accusations.   Wethersfield, CT
"Her confession was attended with
such convictive circumstances that it
could not be slighted." She confessed
that she had murdered a child.
(Cotton Mather, *Magnalia Christi American*, Book 6,7)

# APPENDIX G

**1649**
| | | |
|---|---|---|
| ? Marshfield | witch | Springfield |
| Mary Oliver | confessed witch | Boston |
| | Witchcraft accusations | Windsor, CT |
| (Drake) | | |

**1650**
| | | |
|---|---|---|
| Alice Stratton | midwife accused witch | Watertown |
| (Murphy) | | |
| Jane James | | Marblehead |
| Witchcraft accusations | | Boston |
| (Drake) | | |

**1651**
| | | |
|---|---|---|
| Witchcraft accusations | | Cambridge |
| | | New Haven, CT |
| | | Hartford, CT |
| Joan and John Carrington | nocturnal gatherings | Wethersfield, CT |
| Mary and Hugh Parsons | accused witches | Springfield |
| Sarah Merrick | | |
| Bessie Sewell | | |
| Alice Lake | | Dorchester |
| ? Bassett | accused witch | Stratford? Fairfield, CT |
| (Drake, Taylor) | | |

**1652**
| | | |
|---|---|---|
| Accused witch | | Watertown |
| John Bradstreet | accused witch | Rowley |
| (Drake) | | |

**1653**
| | | |
|---|---|---|
| Witchcraft accusations | | Fairfield, CT |
| | | Lynn |
| | | Gloucester |
| | | New Haven CT |
| (Drake, Demos) | | |

## DISGUISED AS THE DEVIL

### 1654
Witchcraft accusationsWindsor, CT
New Haven, CT

Bystander asked for "Knapps wife to
be new searched after she was
hanged, and when she saw the teates,
said if they were the markes of a
witch, then she was one, or she had such markes"
(Taylor)

Katherine Grady  hanged at sea   weather magicVirginia/
Mary Leehanged at sea   tempestMaryland
(Davis)

### 1655
Witchcraft accusationsIpswich
Elizabeth Goodman, "begger"New Haven, CT
(Drake)

### 1656
Witchcraft accusationsPortsmouth, N.H.
Boston
Springfield
Dover, NH
Salem
Northumberland, VA
Eunice Cole"had foul nasty disposition"Hampton, N.H.
(Davis, Drake)

### 1657
Witchcraft accusationsGloucester
New Haven, CT
MA
(Davis, Drake)VA

### 1658
Elizabeth Garlick   Witchcraft accusationsEasthampton, N.Y
"Goody Garlick is a naughty woman"
Elizabeth Simons, bewitched mother of an infant,
had fever, severe pain, pin found in mouth, fits, died.
(Breen, Lyon)
Elizabeth Richardson  hanged at seaMaryland
Weather magic
Katherine Grade?Hanged at seaVA
(Drake)Weather magic

# APPENDIX G

Sudden illness, death　　　　　　　　　　　　　　Saybrook, CT
(Taylor)

## 1659
Witchcraft accusations　　　　　　　　　　　　Andover
　　　　　　　　　　　　　　　　　　　　　　Saybrook, CT
　　　　　　　　　　　　　　　　　　　　　　Cambridge
　　　　　　　　　　　　　　　　　　　　　　VA
　　　　　　　　　　　　　　　　　　　　　　York (Maine)

(Drake, Taylor)

## 1660
Witchcraft accusations　　　　　　　　　　　　Cambridge
　　　　　　　　　　　　　　　　　　　　　　Scituate
　　　　　　　　　　　　　　　　　　　　　　Wethersfield, CT
　　　　　　　　　　　　　　　　　　　　　　Plymouth
Mary Wright　Quaker heresy　　　　　　　　　Oyster Bay, NY
(Drake, Lyon)

## 1661
Witchcraft accusations　　　　　　　　　　　　Saybrook, CT
(Taylor)
Joan Mitchell accused witch suing for slander　　Charles Co., Maryland
(Parke)

## 1662
Elizabeth Brown　　sensation of bird pricking her　　Salisbury
　　　　　　　　　with motion of…wings..in her
ECASWVI, 63-4　　throat a bunch like a pullet egg

Ann Cole　　Had fits, uttered matter unintelligible　Hartford, Conn.

Daughter of John Kelley　death, red spot on cheek
(*Narratives*, 18, Taylor [*Conn. Col. Rec.*, II: 91], 268-69)

Witch Rebecca Greensmith was "a lewd, ignorant and considerably aged woman"
(Mather's *Magnalia*VI, 71-78)

Witchcraft accusations　　　　　　　　　　　　Haverhill
　　　　　　　　　　　　　　　　　　　　　　Portsmouth, N.H.
　　　　　　　　　　　　　　　　　　　　　　Wethersfield, CT
　　　　　　　　　　　　　　　　　　　　　　Farmington, CT
(Drake)

## DISGUISED AS THE DEVIL

**1663**
Witchcraft accusations					Farmington, CT
								Hartford, CT

(Drake)

**1664**
Mary Hall witchcraft accusations			Setauket, NY
(Lyon)

**1665**
Witchcraft accusations					Haverhill
								Hartford, CT
								Setauket, NY
								Cambridge
								Brookhaven, NY
								St. Marys, Maryland

(Drake, Lyon, Parke)

**1667**
Witchcraft accusations					Marblehead
								Salem
								Stamford, CT
								Saybrook, CT

(Drake)

**1668**
Witchcraft accusations					Wethersfield, CT
Thomas Bracy (probably Tracy)
"was bemoydered in his understandinge
or actinge, Thomas left groninge and
lay quiet a little, and then [the witch]fell
againe to afflictinge and pinching, Thomas
againe groninge…could not speake…
the next day Mr. Marten and his wife saw
the mark of the saide afflictinge and pinchinge."

"Cows [were] runninge with greate violence,
taile on end" "The cattle stood staring and fed not.
In a little time the oxen as affrighted fell to running,
and ran with such violence …The cattle were used
ordinarily before to be so tied and fed--in other places,
 & presently after being so tied on other men's ground
they fed--peaceably as at other times."

# APPENDIX G

Joane Francis' "child continued strangely ill about three weeks, wanting a day, and then died, had fits."

"as he lay in his bed at Windsor in the night he had a great box on the head….. then my husband's nose fell a bleeding in an extraordinary manner, & so continued (if it were meddled with) to his dying day."

Mary Hale reported that "presently something fell on her legs with such violence that she feared it would have broken her legs, and then it came upon her stomach and oppressed her so as if it would have pressed the breath out of her body. Then appeared an ugly shaped thing like a dog…That day seven night next after, lying in her bed something came upon her in like manner as is formerly related, first on her legs & feet & then on her stomach, crushing & oppressing her very sore…presently then she had a great blow on her fingers which pained her 2 days after, which she complained of to her father & mother, & made her fingers black and blue. Mary was again pressed very much"

"one of wch [oxen] spoyled at our stile before our doore, with blows upon the backe and side, so bruised that he was altogether unserviceable; about a fortnight or three weeks after the former, we had a cow spoyled, her back broke and two of her ribs, nextly I had a heifer in my barneyard, my ear mark of wch was cutt out and other ear marks set on; nextly I had a sow that had young pigs ear marked (in the stie) after the same manner; nextly I had a cow at the side of my yard, her jaw bone broke and one of her hoofs and a hole bored in her side, nextly I had a three yeare old heifer in the meadow stuck with knife or some weapon and wounded to death; nextly I had a cow in the street wounded in the bag as she stood before my door, in the street, nextly I had a sow went out into the woods, came home with ears luged and one of her hind legs cutt offe, lastly my corne in Mile Meadow much damnified with horses, they being staked upon it; it was wheat"

Edward Jesop became lost near his home even though he tried to "keep upon my hors & did my best indeauour to get home I was ye greatest part of ye night wandering before I got home altho I was not much more than two miles."
(Taylor)

| | | |
|---|---|---|
| John Pressy | Became disoriented in the woods, | Amesbury, Mass. |
| ECASWVI, 64 | Hallucination? | |

## 1669
Witchcraft accusations

Ipswich
Portsmouth, NH
Hadley
Stamford, CT

Susannah Martin accused    Amesbury
(Drake)

# DISGUISED AS THE DEVIL

**1670**
Witchcraft accusations               Lynn
                                     Wethersfield, CT
(Drake)

**1671**
Elizabeth Knap(p)                    Groton, CT.
Tongue drawn up like a semi circle
to the roof of her mouth, not to be
removed, tongue drawn out of mouth
to extraordinary length...Voice not her own
(Narratives, 21-23)
Witchcraft accusation                Northumberland, VA
(Davis)

**1672**

James Carr                           Salem area

I was taken after a strange manner
as if every living creature did run about
every part of my body...continued for
about three quarters of a year.
(ECASWVII, 38)


Katherine Palmer    accused witch    Newport, Rhode Island
(Demos)

**1673**
Eunice Cole         accused again    Hampton, NH
Anna Edmunds        accused witch    Lynn
? Messenger                          Windsor, CT
Drake, Taylor, Demos)

**1674**
Mary Parsons        witch            North Hampton
(Drake)

**1675**
Witchcraft accusations               Norfolk, VA
(Davis)

# APPENDIX G

**1676**
Mary Ingham          witch                    Scituate
(Demos)


**1677**
Alice Beamon         witch                    Springfield
(Demos)


**1678**
Samuel Gray                                   Salem
Hallucination "between sleeping and waking"
(ECASWVI, 38)

? Burr               accused witch            Wethersfield, CT
(Demos)

**1679**
Morse Boy                                     Newberry, Mass.
Barked like a dog, cluck't like a hen,
in fit his tongue likewise hung out of
his mouth, lay miserable lame, creeping
on one side, lost speech, Grown antic
(Narratives, 23-32)

**1679**
Witchcraft-change in corpse after suspicious death    Accomack Co., VA
(Davis)

**1680**

Shattuck Child                                Salem Town
taken with a drooping condition,
taken in a terrible fit, his mouth and eyes
drawne aside and gasped… at length we
perceived his understanding
decayed –symptoms lasted 17 or 18 months.
(Narratives 225,226,380)

Rachel Fuller        accused witches          Hampton, NH
Isabella Towle
Eunice Cole
Bridget Oliver       accused witch            Salem
Margaret Gifford                              Lynn
(Demos)

247

## DISGUISED AS THE DEVIL

## 1681

Mary Hale        accused witch        Boston
(Demos)

## 1682

Bernard Peach        Salisbury
"one of the best cows...was in such a
mad fright that two men had much
ado to get her...she did run and fly about."
(ECASWVI, 68)

Hannah Perley        Ipswich
" was so sore, fell in a dreadful fit,
complaining of being pricked with
pins and falling down into dreadful
fits, so pined a wai to skin and bone and
ended her sorrowful life." Also, had a
cow "taken strangely running about
like a mad thing...."
(ECASWVI, 64,146)

Howen Boy        Boston area
 Taken in same way as Goodwin children (see 1688)
(Narratives, 37)

?        accused witch        Kittery, Maine
?        accused witch        Hartford, CT
(Demos)

## 1683

Witchcraft accusations        Hartford, CT
Mary Hortado        Salmon Falls, ME
"bitten on both arms black and
blue and one of her breasts scratched, the
impression of the teeth being like a mans
Teeth...were plainly seen by many."
(Narratives, 36)

James Fuller     accused        Springfield
? Travally     accused        South Hampton, NY
Mary Webster     accused        Hadley
(Demos, Drake)

# APPENDIX G

## 1684

John Louder                                                            Salem area
felt as if a great weight were
pressing on body, struck dumb
(ECASWVI, 41)

Philip Smith                                                      Hadley
The jury that viewed his corpse
found a swelling on one breast,
his privates wounded or burned,
his back full of bruises, and several
holes that seemed made with awls. (Lyme co-infection? Petechiae is associated with babesiosis )

(Mather, Vol II Chapter 7)

Witchcraft accusations- two old Swedish women and Quakers     Pennsylvania
(Robbins)

## 1685

Shattuck child    (See 1680)                                Salem Town
"was taken in a strange and unusual
manner as if his vitals would have
broke out, his breastbone drawn up
togather to his upper part of his brest,
His neck and ey(e)s drawne so aside
as if they would never come to right
Againe, (is) generally in such an uneasie
and restless frame almost always
Running to and fro acting so strange."

John Roger                                                        Billerica
a cow "of a sudden she would give little or none [milk]"
(ECASWVI, 139)

Jarvis Ring                                                       Salisbury
felt "something coming upon him when he
was in bed and did sorely afflict him by lying
upon him and he could neither move nor speake
while it was upon him...the print of the bite
(of a woman) is yet to be seen on the finger of
his right hand for it was hard to heal"
(Narratives, 235)

 **DISGUISED AS THE DEVIL**

Mary Webster             accused of witchcraft                        Hadley
(Demos)

Witchcraft accusations                        St. Mary's Co., Maryland
(Parke)

## 1686

Goodwife Trask                        Salem bounds Bordering Beverly
Distracted, had fits that were the
same as the afflicted girls,
may have committed suicide by
slashing throat with scissors
(ECASWVI, 96-7)

Witchcraft accusations                        St. Mary's Co., Maryland

## 1688

The Godwin Children                                        Boston
Bewitched, strange fits, seized sometimes deaf, dumb, or blind, tongues down throat, pulled out upon chins to prodigious length, mouths opened to such a wideness that their jaws went out of joint, shoulder blades, elbows, wrist, several of joints stiff, no stirring of head, heads would be almost twisted around, lost hearing, complain that they were in a hot oven, red streaks on body, could not move head, stiff then limber, they would be twisted into such postures…
(Narratives, 99-128)

## 1689

? Bowden                accused witch                        New Haven, CT
(Demos)

Benjamin Holton                                        Salem Village
Had fits that were the same as the
afflicted girls, was much pained at his
stomach and was often struck blind. Died.

Sam. Abbey                                        Salem Village
Lost several cows in drooping condition,
had a cow that could not rise alone
(ECASWVI, 8)

## APPENDIX G

Thomas Gadge ?
Cows died in sudden, terrible & strange unusual manner
(ECASWVI, 8)

Margaret Reddington Salem?
 Hallucination, I was exedingly elle
(ECASWVI, 125)

## 1690
Priscilla Stacey ?
felt "pinching and bruising of her til
earms and other parts of her body
Looked black by reson of her soer
 pinching of her in the time of her sickness.
Died.
(ECASWVI, 37)

## 1691
Benjamin Abbott Andover
I was taken with a swelling in my foot
& then was taken with a pain in my side
which bred to a sore...some of the cattle
would come in from the woods with their
tongues hanging out of their mouths in a
strange and affrightening manner.
(ECASWVI, 138)

Mary Hale     accused witchcraft     Boston

Mary Randall  accused witchcraft     Northampton

(Demos)

William Beale Marblehead
A great and wracking pain
Hallucination of shade or shape...Nosebleed
(Special Collections
New York Public Library)

Sam Wilkins Salem Village
I was also pinced by an unseen hand
Daniel Wilkins
Fever, died, had red marks like stabs of awl on body (Lyme co-infection? Petechiae is associated with babesiosis)
 (ECASWVI, 101)

251

 DISGUISED AS THE DEVIL

## 1692
## CONNECTICUT OUTBREAK

| | | |
|---|---|---|
| Mercy Disborough | accused witch | Fairfield, CT |
| Elizabeth Clawson | accused witch | Stamford, CT |
| Mary Staples | accused witch | Fairfield, CT |
| Mary Harvey | accused witch | Fairfield, CT |
| Hannah Harvey | accused witch | |
| ? Miller | accused witch | |
| Winifred Benham | accused witch | Wallingford, CT |
| Hugh Croasia | accused witch | Stratford, CT |

John Barlow " I was in bedd ...Marsey Desbory came to me and layed hold on my fett and pinshed them (and) looked wishley in my feass and I strouff to rise and cold not and too speek and cold not.."

Henry Grey had a "calfe very strangly taken oared very strangly for ye space of near six or seven howers & also scowered extraordinarily all which after an unwonted maner; & also saith he had a lame after a very strange maner it being well and ded in about an houre and when it was skined it lookt as if it had been bruised orpinched on ye shoulders"

John Grummon senr "saith yt about six year agou he being at Compo with his wife & child & ye child being very well as to ye outward vew and it being suddenly taken very ill & so remained a little while upon wch he being much troubled went out & heard young Thomas Benit [his nephew] threaten Mercy Disbrow & bad her unbewitch his uncles child."

On September 15, 1692 Daniel Westcott's "gerle"[servant] Cateron Branch "when first ye garl was taken with strang fits she was sent for to Danil Wescots house & she found ye garle lieing upon ye bed. She then did apprehend yt the garls illness might be from sum naturall cause; she therefore adnised them to burn feathers under her nose & other menes yt had dun good in fainting fits and then she seemed to be better with it; and so she left her that night in hops to here she wold be better ye next morning; but in ye morning Danil Wescot came for her againe and when she came she found ye garl in bed seemingly senceless & spechless; her eyes half shet but her pulse seemed to beat after ye ordinary maner her mistres desired she might be let blud on ye foot in hops it might do her good."

"The deponant saith that when the garl put her hand ouer the bed it was open and he looked very well in her hand and cold see nothing and before shee puled in her hand again shee had goten yt pin yt hee took from her."

"she had a violent fit and calling again said now they are agoing to killme & crieing out very loud that they pincht her on ye neck and calling out yt they pincht her again I setting by her I took ye light and look upon her neck & I see a spot look red seeming to me as big as a pece of eight afterwards it turned blue & blacker then any other part of her skin and after ye second time of her calling I took ye light & looked again and she pointed with her hand lower upon her shoulder and I se another place upon her shoulder

## APPENDIX G

look red & blue as I saw upon the other place before and then after yt she had another fit."

"when she was in those fits ratling in her throat she would put out her tong to a great extent I consieue beyond nature & I put her tong into her mouth again & then I looked in her mouth & could se no tong but as if it were a lump of flesh down her throat and this ofen times."

"Being in ye feilds gathering of herbs, she was seizd with a pinching & pricking at her breast; she being come home fell a crying, was askd ye reason, gave no answer but wept & immediately fell down on ye flooer wth her hands claspt, & with like actions continued wth some respite at times ye space of two days"

In Novembor 1692 " Hugh Crosia is complained of by a gerll at Stratford for aflicting her ....undar suspecion offamiliarity with satan sd Crosha being asked whethar he sayd he sent ye deuell to hold downe Eben Booths gerll"

(Taylor)
## SALEM OUTBREAK

Dorcas GoodHomeless? Salem area
accused witch had a deep red spot
about the bigness of a fleabite. Afflicted for life.
(ECASWV2, 67)

Sarah GoodHomeless?Salem area
Foul argumentative disposition, muttered to Self.
(ECASWV1,10)

Phoebe ChandlerAndover
There was a voice in the bushes...
one half of my right hand was greatly
swollen and exceedingly painful& lso
part of my face...several times
since I have been troubled with a
Weight upon my breat & upon my legs
Soe that I could hardly go..I had a Strange
burning in my stomach & then
was struck deaf
(ECASWVI, 140)

Ann Putnam Sr.
Sickness, fatigue, fits, eventually dies.Salem Village
Ann Putnam Jr.
Eldest daughter also taken in a sad manner
(Narratives, 157)

253

## DISGUISED AS THE DEVIL

John How ?
Sow died. Hand so numb and full of pain .....
it is not wholly well now
(ECASWVI, 151)

John Parker                              Andover ?
Josiah Eaton
He was pressed almost choked..
his son was strangely handled and sometimes
blind in one ey and sometimes in the other ey
(Suffolk Court Records Case 2710, 48)

Stephen Bittford ?
I was in grate pain of my neck and could
not stir my head nor spake a word. I r'cvd
Blow on my chest I was piched and nipt
By sumthing invisible for sum time.

Abraham Wellman ?
Cow was taken with fits...when she saw
any person coming to milk her she would
run and let no one near her for a week
(Suffolk Court Records Case 2712, 48)

Benjamin Gould ?
I had two penches upon my side..I had
then such a paine in one of my feet that
I could not ware my shoe for 2 or 3 days.
(ECASWVII, 42)

Giles Corey                              South of Salem Village
I fetched an ox out of the woods..he laying
down in the yard..I went to raise him to yoke
him but he could not rise up but dragged his
hinter parts as if he had been hiptshott. I had a
cat last week strangely taken.
(ECASWV1, 13)
Martha Corey          Muttered, talked to self
See Giles above

Dog                                      Salem Village
afflicted, put to death
 (Narratives, 372)

# APPENDIX G

| | | |
|---|---|---|
| Mary Parker(accused witch) | | Andover |

Mrs. Parker did lay upon the durt and snow…
they thought she was dead
a neighbor said she saw her before in
Such kind of fits…she rises up and laughs in
or faces…I have seen Mrs. Parker in such
condition several other Times.[there is a folkloric
belief that when a witch transforms into a demon
she leaves behind her body in a trance like state]
(ECASWV2,33)

| | | |
|---|---|---|
| Mercy Short (Narratives, 264) | fits, pins on body | Charlestown |
| Elizabeth Parris (Narratives,160) | afflicted girl, fits | Salem Village |
| Elizabeth Hubbard (ECASWVI, 9) | afflicted girl...they "afflict me by pinching and pricking me" | Salem Village |
| Mary Walcut (Narratives, 53) | had marks of teeth on wrist | Salem Village |
| Abigail Williams (Narratives, 153) | had mark like the print of an orange on her skin | Salem Village |

## 1693
Witchcraft accusations    Stratford, CT
                          Andover

(Demos)

## 1694
| | | |
|---|---|---|
| Witchcraft accusations | press of weight on chest | King &Queen Co. VA |
| Witchcraft accusations | sick horse | Westmoreland, VA |

(Davis)

## 1695
Witchcraft accusations    Westmoreland, VA
                          King &Queen Co, VA

(Davis)

## 1697
| | | |
|---|---|---|
| Winifred Benham | accused of witchcraft, frequently and sorely afflicted in the bodies of victims | Wallingford, CT |

Winifred Benham, Jr
(Drake)

## DISGUISED AS THE DEVIL

Witchcraft accusations                                Norfolk, VA
(Davis)

**1702**
Slander/witch accusations                             Maryland
(Parke)

**1703**
Judge Nichols Trott of Charleston accused        S. Carolina
(Robbins)

A witch was accused of mixing up words while speaking     Virginia
(Davis)

**1724**
Elizabeth Ackley accused Sarah Spencer of          Colchester, CT
"riding and pinching." Sarah sued
for 500 lbs. damages and got
judgment for 5,with costs. The
Ackleys appealed, and at the trial
the jury awarded Sarah damages
of 1 shilling, and also stated that
they found the Ackleys not insane.
(Taylor)

**1768**                                                      Bristol, CT
"On the mountain," a young woman named Norton accused her aunt of putting a bridle on her and driving her through the air to witch meetings in Albany, caused a commotion. Deacon Dutton's ox was torn apart by an invisible agent, and unseen hands brought new ailments to the residents there, pinched them and stuck red hot pins into them." [By this time complaints have moved away from settled areas into the mountains of Connecticut.]
(Taylor)

### Sources and Abbreviations for Appendix G:
**Bradford-** William Bradford. *Of Plymouth Plantation 1620-1647.* New York: The Modern Library, 1981
**Breen-**Breen, T.H. *Imagining the Past: East Hampton Histories.* Athens, Georgia: University of Georgia Press, 1996.
**Davis-**Richard Beale Davis, "The Devil in Virginia in the Seventeenth Century" *Virginia Magazine of History and Biography* 65 (1957), 131-49.
**Demos-**John Putnam Demos. *Entertaining Satan: Witchcraft and the Culture of Early New England.* New York: Oxford University Press, 1983 (Appendix A, 402-9).

# APPENDIX G

**Drake-**Frederick Drake, "How North Carolina Juries Dealt with Charges of Witchcraft", *North Carolina Historical and Genealogical Register* 2 (April, 1901) 56-59, 68-69.

**ECASWVI or II-** Essex County Massachusetts Archives Salem Witchcraft, Vol. I and II.

**Krech-** Shepard Krech, *The Ecological Indian, Myth and History*. New York: W.W. Norton & Co., 1999.

**Lewis-** Lewis, B.A. "Prehistoric juvenile rheumatoid arthritis in a pre-contact Louisiana native population reconsidered,"*American Journal PhysicalAnthropology,*104 June 1998.

**Lyon-** John Lyon "Witch craft in New York" *New York Historical Collections* 2(1869), 273-6.

**Mather-** Cotton Mather, *Magnalia Christi American,* (Cheapside, London: Thomas Parkhurst, 1702).

**Murphy-** Edith Murphy, "Skillful Women and Jurymen: Gender and Authority in Seventeenth-Century Middlesex County, Massachusetts" (unpublished Ph.D. dissertation, University of New Hampshire, 1998).

**Narratives-**Narratives of the New England Witchcraft Cases, edited by George Lincoln Burr (Mineola, New York: Dover, 2002).

**Parke-**Francis Parke, "Witchcraft in Maryland," *Maryland Historical Magazine* 31:4(December 1936) 271-298.

**Robbins-**Russell H. Robbins. *The Encyclopedia of Witchcraft and Demonology.* New York: Bonanza Books, 1959.

**Rothschild-**Bruce M. Rothschild, "Tennessee Origins of Rheumatoid Arthritis" *McClung Museum Research Notes,* No.5, April 1991.

**Schutte-**Anne Jacobson Schutte, "'Such Monstrous Births': A Neglected Aspect of the Antinomian Controversy," *Renaissance Quarterly* 38 (1985): 85–106.

**Smith-** John Smith. General Historie of Virginia, (1624).

**Suffolk Court Records**, Massachusetts.

**Taylor-** John M. Taylor, *The Witchcraft Delusion In Colonial Connecticut 1647-1697* (1908) electronic edition available online at www.fullbooks.org.

**Winslow-** Edward Winslow, *Mourt's Relation.* Bedford, Massachusetts: Applewood Books, 1963.

**Winthrop-** John Winthrop, *Journal*

---

# APPENDIX G

[579] Spanish colonies in the New World continued Spain's history of little interest in witch hunting although New Mexico, a Spanish colonial settlement, saw a few witchcraft accusations which appear to be related to Native American shamanism. It is, however, not a modern *Ixodes scapularis* endemic area Surprisingly French Canada was also fairly free of witch-hunts, even though it was popular back home in France. This may be because Canada was too cold during the Little Ice Age for an *Ixodes* population to be well established, although that is now changing. See Paul Nance III, "Tale of the Tick: How Lyme disease is expanding Northward," *Northern Woodlands* Issue 56 (Spring 2008) for an excellent article. Descriptions included in this list from Connecticut are lengthier because they were not used within the text as extensively as those from Massachusetts but are still highly relevant. Descriptions have been left in 17th century dialect as much as practical.

# DISGUISED AS THE DEVIL

# APPENDIX H

**Changing Population Levels Altered the Environment and Landscape**

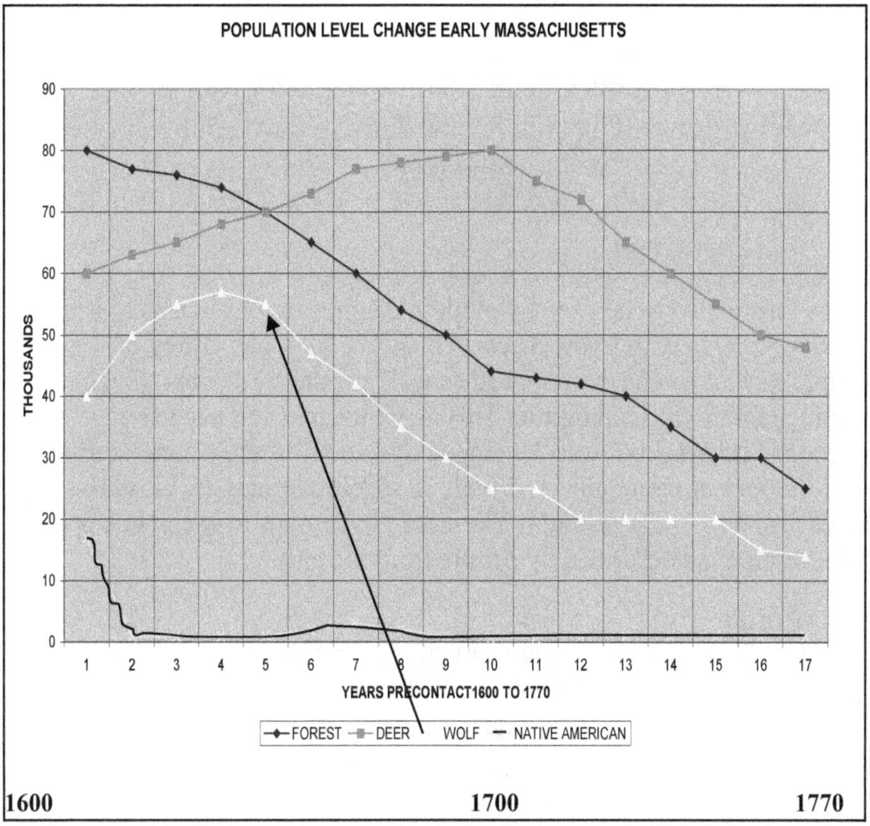

POPULATION LEVEL CHANGE EARLY MASSACHUSETTS

[Graph showing population levels (in thousands) of Forest, Deer, Wolf, and Native American from years precontact 1600 to 1770]

1600 — 1700 — 1770

**1600-30:** Epidemic sickness among Native Americans, population drops to 900 by the time English settlers arrive, may be extension of plague pan-epidemic from Europe.
**Before 1640:** Bounties introduced, wolf population begins decline.
**1670's:** King Philip's War, followed by Native American population decline.
**1680's-1692:** Drought affects humans and animals.
**1690's:** King William's War
**After 1690-** Improved rifled muskets begin to become available, starts decline of deer as hunting becomes instituted in culture.
**Forests-**Deforestation continues steadily throughout the time period shown.

 DISGUISED AS THE DEVIL

APPENDIX J

# LYME DISEASE[580]

## Symptoms Check List, Diagnosis Criteria, and Definition of Chronic Lyme disease that can be applied to the past.

SYMPTOM CHECK LIST
Note: This is a checklist that was given to me by one of my treating physicians.
It was formulated to be used by modern diagnosing physicians to help identify patients with Lyme disease. It shows the wide variety of symptoms that can be experienced as well as other problems (like heart murmur) that Lyme can mimic. This is not meant to be used as a diagnostic scheme, but was developed to streamline the office interview. Note the format-complaints referable to specific organs are clustered to better display multi-system involvement. Some parts of it can be applied to the symptoms described by people from the past.

RISK PROFILE:
Tick infested area___ Frequent outdoor activities ___Hiking___
Fishing___Camping____Gardening___Hunting___
Ticks on pets__
Other household members with Lyme___
Do you remember being bitten by a tick?.....N0___YES___when___
Do you remember having the 'Bull's Eye Rash'? NO___YES___
Any other rash? NO____YES_____

Have you had any of the following? CIRCLE ALL YES ANSWERS
1. Unexplained fevers, sweats, chills or flushing
2. Unexplained weight change-(loss or gain)
3. Fatigue, tiredness, poor stamina
4. unexplained hair loss
5. Swollen glands: list areas_____
6. Sore throat
7. Testicular pain/pelvic pain
8. Unexplained menstrual irregularity
9. Unexplained milk production, breast pain
10. Irritable bladder or bladder dysfunction
11. Sexual dysfunction or loss of libido
12. Upset stomach or abdominal pain
13. Liver problems

# APPENDIX J

14. Change in bowel function (diarrhea, constipation)
15. Chest pain or rib soreness
16. Shortness of breathe, cough
17. Heart palpitations, pulse skips, heart block
18. Any history of heart murmur or valve prolapse?
19. Joint pain or swelling: list joints_____
20. Stiffness of the joints or back
21. Muscle pain or cramps
22. Twitching of the face or other muscles
23. Headache
24. Neck creaks and cracks, neck stiffness, neck pain
25. tingling, numbness, burning or stabbing sensations, shooting pains, skin hypersensitivity
26. Facial paralysis (Bell's Palsy)
27. Eyes/vision: double, blurry, increased floaters, light sensitivity
28. Ears/ hearing: buzzing, ringing, ear pain, sound sensitivity
29. Increased motion sickness, vertigo, poor balance
30. Light headedness, wooziness, unavoidable need to sit or lay down
31. Tremor
32. Confusion, difficulty in thinking
33. Difficulty with concentration, reading
34. Forgetfulness, poor short term memory, poor attention, problem absorbing new information
35. Disorientation: getting lost, going to wrong places
36. Difficulty with speech or writing; word or name block
37. Mood swings, irritability, depression
38. Disturbed sleep-too much, too little, fractionated, early wakening
39. Exaggerated symptoms or worse hangover from alcohol

DIAGNOSTIC CHECKLIST

To aid the clinician, a workable set of diagnostic criteria were developed with the input of dozens of front line physicians. The resultant document has proven to be extremely useful not only to the clinician, but it can also help clarify the diagnosis for third party payers and utilization review committees. It is important to note that the CDC's published guidelines are for surveillance only, not for diagnosis.

| LYME BORRELIOSIS DIAGNOSTIC CRITERIA | VALUE |
|---|---|
| {that can be generally applied to situations in the historical past} | |
| Tick exposure in endemic area | 1 |
| Historical facts and symptoms consistent with Lyme | 2 |
| Symptomatic signs & symptoms consistent with Bb infection | |
|     Single system, e.g. monoarthritis | 1 |
|     Two or more systems, e.g., Monoarthritis/facial palsy | 2 |
| ECM, physician confirmed | 7 |

DIAGNOSIS[581]: Add together criteria values-

| | |
|---|---|
| Lyme borreliosis Highly Likely | 7 or above |
| Lyme borreliosis Possible | 5-6 |
| Lyme borreliosis Unlikely | 4 or below |

CHRONIC LYME DEFINITION:

To be said to have chronic Lyme, these three criteria must be present:

1. Illness present for at least one year
2. Have persistent major neurologic involvement (such as encephalitis/encephalopathy, meningitis, etc.) or active arthritic manifestations (active synovitis)
3. Still have active infection with *B. burgdorferi*, regardless of prior antibiotic therapy (if any).

---

[580]See Karen Vanderhoof-Forschner, *Everything You Need to Know about Lyme disease and other tick-borne disorders*( New York: John Wiley & Sons, Inc., 1997), Denise Lang, *Coping with Lyme disease* (New York: Holt, 2004)and Dr. Joseph Burrascano, Jr., *Diagnostic Hints and Treatment Guidelines for Lyme and other tick borne illnesses*, 14th Edition, November, 2002, at www.ilads.org.

[581] **Some of the afflicted** girls of 1692 had "bite marks" as well as other symptoms while living in a Lyme endemic area, if the marks were ECM's they would be placed in the Lyme borreliosis highly likely category.

# SELECTED BIBLIOGRAPHY

Anderson, Virginia DeJohn. *New England's Generation: The Great Migration and the Formation of Society and Culture in the Seventeenth Century.* Cambridge, U.K.: Cambridge University Press, 1991.
_____.*Creatures of Empire.* New York: Oxford University Press, 2004.
Aronowitz, Robert. *Making Sense of Illness, Science, Society and Disease.* Cambridge U.K.: Cambridge University Press, 1999.
Atkins, Robert. *Dr. Atkins New Diet Revolution.* New York: Avon Books, 2002.
Arnold, Janet. *Patterns of Fashion: The Cut and Construction of Clothes for Men and Women c.1560-1620.* U.K.: Drama Publishers, 1985.
Bailey, M.D. "The Disenchantment of Magic: Spells, Charms, and Superstition in Early European Witchcraft Literature." *The American Historical Review* Vol. 3 No. 2, April, 2006, available online at www.historycooperative.org.
Baker, Emerson W. and John Reid. *The New England Knight: Sir William Phipps, 1651-1695.* Toronto: University of Toronto Press, 1998.
Baker, James. '17th Century Yeoman Foodways." *in The Plimouth Plantation Cookbook.* Massachusetts: Plimouth Plantation, 1990.
Barbour, Alan and Durland Fish. "The biological and social phenomenon of Lyme disease." *Science* 260, June 11, 1993, 1610-1616.
Barbour, Philip, ed., John Smith, *General Historie of Virginia, New England and the Summer Isles,*(London, 1624) in *Complete Works of Captain John Smith.* Chapel Hill: University of North Carolina Press, 1986.
Beck, Louisa, *CHAART Lyme disease New York- Municipality Study of Lyme disease.* July 2000 at http://geo.arc.nasa.gov.
Beck, Louisa, Bradley Lobitz and Byron Wood, "Remote Sensing and Human Health: New Sensors and New Opportunities." *Emerging Infectious Diseases* 6 No. 3, April 2000.
Behringer, Wolfgang. *Witches and Witch-Hunts.* Malden, Mass.: Polity, 2006.
Behringer, Wolfgang. "Climatic change and Witch-Hunting: The Impact of the Little Ice Age on Mentalities." *Climatic Change,* 43,1999, 335-51.
Ben-Yehuda, N. "Problems Inherent in Socio-Historical Approaches to the European Witch Craze." *The Journal for the Scientific Study of Religion,* 20(4), 1968, 326-338.
Bergesen, A. A Durkheimian Theory of "Witch-Hunts" with the Chinese Cultural Revolution as an Example, *The Journal for the Scientific Study of Religion,17 (1)* 1978, 19-29.
Bergesen, A. *The Sacred and the Subversive: Political Witch Hunts as National Rituals,* Monograph Number 4. Storrs, Ct: The Society for the Scientific Study of Religion, 1984.
Bleiweiss, John. "When to suspect Lyme." from www.cassia.org/essay.htm.
Bodin, Jean. *On the Demon-Mania of Witches,* 1580
Boyer, Paul and Stephen Nissenbaum, eds. *Salem Possessed: the Social Origins of Witchcraft.* Cambridge, Massachusetts: Harvard University Press, 1974.
_____ *Salem-Village Witchcraft: A Documentary Record of Local Conflict in Colonial New England.* Boston: Northeastern University Press, 1993.

Bradford, William. *Of Plymouth Plantation 1620-1647*. New York: The Modern Library, 1981.
Briggs, Robin *Witches and Neighbors: The Social and Cultural Context of European Witchcraft*. New York: Viking, 1996.
Brorson, O. and S, Brorson, "A Rapid Method for Generating Cystic forms of *Borrelia burgdorferi* and their reversal." *APMIS* 106 No.12, December 1998.
Burgdorfer, Willy. Keynote address "The complexity of Vector-borne spirochetes" given on April 9, 1999 at the *Twelfth International Conference on Lyme Disease and Other Spirochetal and Tick borne Diseases*.
Burke, G., et al., "Hypersensitivity to Ticks and Lyme disease Risk in Heavily Tick Infested Areas." *Journal of Emerging Infectious Diseases* 11 No.1, January 2005.
Burr, George Lincoln. *Narratives of the New England Witchcraft Cases*. Mineola, New York: Dover Publications, 2002.
Bushmich, S. "Lyme borreliosis in domestic animals." *Journal of Spirochetal and Tick-borne Diseases*, Vol. 1, No. 1, 1994, 24-28.
Cairns, Conrad. *Medieval Castles*. Cambridge, U.K.: Cambridge University Press, 1987.
California Department of Environmental Science Policy and Management, *Wild Pigs*. Berkeley, California: California Printing Office, 1998.
Caporael, Linnda. "Ergotism: The Satan Loosed in Salem?" *Science* 192, 1976.
Carlson, Laurie Winn. *A Fever in Salem*. Chicago: Ivan R. Dee, 1999.
Carroll, Michael Christopher. *Lab 257, The Disturbing Story of the Government's Secret Plum Island Laboratory*. New York: Morrow, 2004.
Carroll, John. "Kairomonal activity of white tailed deer metatarsal gland substance: a more sensitive bioassay using Ixodes scapularis." *Journal of Medical Entomology* 35, 1996.
_____, "Relative Potential for Acquiring Nymphs of Ixodes Scapularis while walking, crawling or sitting in a Deciduous Woodland." TEKTRAN *United States Department of Agriculture- Agriculture Research Service News*. September 10, 1999.
Center for Disease Control, *Lyme Disease*. Atlanta: U.S. Publishing Office, 2001.
_____, *Lyme Incidence Report*. 2006.
Chase, Kenneth. *Firearms: A Global History to 1700*. New York: Cambridge University Press, 2003.
Chartier, Craig S. *Livestock of Plymouth colony 1620-1692 research report*, n.p. for Plymouth Archaeology Rediscovery.
Childs, J. "Shared vector zoonoses of the Old World and New World: Homegrown or Translocated." *A Congressional Report*. Atlanta, Georgia: United States Center for Disease Control and Prevention, 1998, 1095-1105.
Christianson, Eric. *Sickness and Health in America*. Madison: University of Wisconsin Press, 1997.
Columbus, Christopher. Letter to the King and Queen of Spain, November of 1502. *Christopher Columbus Papers*. National Archives of Spain.
Conti, Richard P. "The Psychology of False Confessions." *The Journal of Credibility Assessment and Witness Psychology* 2 No.1, 1999.

# SELECTED BIBLIOGRAPHY

Cotta, John. *The Trial of Witchcraft Showing the True and Right Method for the discovery with A Confusion of Erroneous Wayes,* London: George Purslowe, 1616.

Cotton, John. Democracy as a Detriment to Church and State *in Essential Documents in American History* (1636) at Academic search Premier http://ebscohost.com.

Cowley, Geoffrey and Anne Underwood, "A Disease in Disguise: Lyme can masquerade as migraine, or as madness." *Newsweek,* August 24, 2004

Cressy, David. *Coming Over.* Cambridge U.K.: Cambridge University Press, 1987.

Cronin, William. *Changes in the Land: Indians, Colonists and the Ecology of New England.* New York: Hill and Wang, 1984.

Crosby, Alfred. *The Biological Expansion of Europe 900-1900.* Cambridge U.K.: Cambridge University Press, 1993.

_____.*The Columbian Exchange: Biological and Cultural Consequences of 1492.* New York, Praeger Publishing, 2003.

Daneau, Lambert. *A Dialogue of Witches.* London, 1575.

Davis, Richard Beale. "The Devil in Virginia in the Seventeenth Century." *Virginia Magazine of History and Biography* 65 (1957), 131-49.

Deetz, James and Patricia Scott Deetz. *The Times of Their Lives.* New York: Anchor Books, 2000.

Del Rio, Martin. *Disquisitiones Magicarum,* 1603.

Demos, John. "Underlying Themes in the witchcraft of 17th Century New England." *American Historical Review* 75, 1970, 1311-1326.

_____ *Entertaining Satan-Witchcraft and the Culture of Early New England.* New York: Oxford Press, 1983.

_____. *A Little Commonwealth: Family Life in Plymouth Colony.* (New York: Oxford University Press, 1999.

Dister, S.W., D. Fish, S.M. Bros, D.H. Frank and B.L. Wood, "Landscape characterization of peridomestic risk for Lyme disease using satellite imagery." *American Journal of Trop. Med. Hyg.* 57 No.6 December, 1997.

Donahue, Brian. *The Great Meadow: Farmers and the Land in Colonial Concord.* New Haven: Yale University Press, 2007

Dow, George. *Everyday life in Massachusetts Bay Colony.* New York: Blom Publishers, 1967.

Durkeim, E. *Division of Labor in Society.* Free Press: Reprint Edition, 1997.

_____ *The Elementary Forms of the Religious Life.* Free Press: Reprint Edition, 1995.

Earle, Carville "Environment, Disease and Mortality in Early Virginia" in *The Chesapeake in the Seventeenth Century.* Chapel Hill, N.C.: University of North Carolina Press, 1979.

Elliot, Daniel, Stephen Eppes, Joel Klein. "Teratogen Update: Lyme disease. "*Teratology,* 2001, 276-281.

Emerson, Everett ed. *Letters from Massachusetts Bay Colony 1629-1638.* Amherst: University of Massachusetts press, 1976.

Erickson, K. T. *Wayward Puritans: A Study in the Sociology of Deviance*, New York: Wiley, 2004.

Essex County, Massachusetts, Archives, *Salem Witch Trials,* Volumes I, II, and III.

Fagan, Brian. *The Little Ice Age.* New York: Basic Books, 2000.

Fallon, Brian and Jenifer A. Mields, "Lyme Disease: A Neuropsychiatric Illness." *American Journal Psychiatry* 151 No.11, November 1994.

Fearnley-Whittingstall, Jane. *Ivies*. London: Chatto & Windus Limited, 1992.

Feder, H. and M. Hint. "Pitfalls in the Diagnosis and Treatment of Lyme disease in Children." *Journal of the American Medical Association*. July, 1995, 66-68.

Fischer, David Hackett. *Albion's Seed: Four British Folkways in America.* New York: Oxford University Press, 1991.

Fitzhugh, William and Elizabeth Ward. *Vikings, The North Atlantic Saga.* Washington, D.C.: Smithsonian Institute Press, 2000.

Fontaine, Jacques. *Discours des marques des sorcier et de la réelle possession que le diable prend sur le corps des hommes.* Lyons, 1611; Arras, [c. 1850])

Frank, Christina, Alan Fix, Cesar Pena, and G. Thomas Strickland, "Mapping Lyme Disease Incidence for Diagnostic and Preventive Decisions, Maryland." *Emerging Infectious Diseases* 8 No. 4, April 2002.

Franzen, Harald. "Ancient Tick Poses New Questions." *Scientific American.* March 28, 2001.

Fraser, Claire M., et al., "Genome sequence of a Lyme disease spirochete, *Borrelia Burgdorferi*." *Nature* 390, December 11, 1997.

Gardner, T. "Lyme Disease" in *Infectious Diseases of the Fetus and Newborn.* New York: Remington Saunders, 1995.

Greiner, L. "Evolution and Revolution as Organizations Grow," *Harvard Business Review.* May-June, 1998, 55-67.

Gevitz, Norman. "'The Devil Hath Laughed at the Physicians': Witchcraft and Medical Practice in Seventeenth-Century New England." *Journal of the History of Medicine* Vol. 55, January 2000.

Ginsberg, Howard S., ed. *Ecology and Environmental Management of Lyme Disease*. New Brunswick, N.J: Rutgers University Press, 1993.

Glass, G.E., et al., "Environmental risk factors for Lyme disease identified with Geographic Information Systems." *American Journal of Public Health* 85 No. 7 July 1995.

Goodhart, Lawrence B. "The distinction between witchcraft and madness in Colonial Connecticut." *History of Psychiatry,* Vol. 13, 2002, 433-444.

Gookin, Daniel. "Historical Collections of the Indians of New England" *Archive* at the Massachusetts Historical Society, 1674, 7-12.

Gordis, Leon. *Epidemiology.* New York: W. B. Saunders Company, 2000.

Grann, David. "Stalking Dr. Steere Over Lyme disease," *The New York Times Magazine*. June 17, 2001 archives at www.nytimes.com.

Greiner, L. Evolution and Revolution as Organizations Grow, *Harvard Business Review.* May-June, 1998, 55-67.

Guazzo, Francesco M. *Compedium Maleficarum,* Milan, 1608.

# SELECTED BIBLIOGRAPHY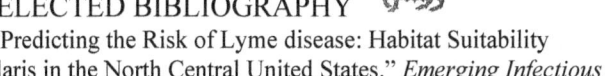

Guerre, Marta, et al., "Predicting the Risk of Lyme disease: Habitat Suitability for Ixodes scapularis in the North Central United States." *Emerging Infectious Diseases* 8, No.3 March, 2002.

Guide to the National Archives of the United States. *Revolutionary War Pension Applications.* Washington, D.C.: National Archives and Records Service, General Services Administration, 1987.

_____.*Civil War Veteran Census, 1890.* Washington, D.C.: National Archives and Records Service, General Services Administration, 1987.

Gylfe, A., et al., "Reactivation of Borrelia infection in birds," *Nature* 403 February 17, 2000.

Hall, A.K., G. Jones and H.K. Kenward, "Cereal bran and human faecal remains from archaeological deposits- some preliminary observations." *Environment and Economy- British Archeological Reports,* 1983.

Hambrick-Stowe, C. "Ordering Their Private World." *Christianity Today International,* Issue 41, May 1996, Vol XIII, No.1, p 16.

Hammond, William. *General Considerations for Planting New England: Chronicles of the colony of Massachusetts Bay 1623-1636.*

Harvey, B. ed., *Living and Dying in England 1100-1540.*Oxford, U.K.: Clarendon Press, 1993.

Harvey, W. and P. Salvato. "Lyme disease: Ancient Engine of an unrecognized *Borreliosis* pandemic." *Medical Hypothesis.* 2003, 742-759.

Hill, Frances. *Hunting for Witches.* Beverly, Mass.: Commonwealth Editions, 2002.

Hessl, A.E. and W.L. Baker, "Spruce and Fir regeneration and climate in the forest tundra ecotone of Rocky Mountain National Park, Colorado." *Arctic and Alpine Research* 29, 1997.

Hinnebusch and A. G. Barbour. "Linear plasmids of *Borrelia burgdorferi* have a telomeric structure and sequence similar to those of a eukaryotic virus." Journal *Bacteriology* 173 No.22, 1991, 7233-7239.

Hoskinson, R.L., and L.D. Mech. 1976. "White-tailed deer migration and its role in wolf predation." *Journal of Wildlife Management* 40, 429-441.

Huff, Mel Symposium: Lyme disease spreading worldwide, *Times Argus Manchester,* August 5, 2007

Hume, Noel Ivor. A Guide to Artifacts of Colonial America. Philadelphia: University of Pennsylvania Press, 1969.

Internet Medieval Sourcebook at www.Fordham.edu

James, Jr., Sydney, Ed. *Three Visitors to Early Plymouth: Letters about the Pilgrim Settlement in New England during its First Seven Years.* Bedford, Massachusetts: Applewood Books, 1963.

Johnson, Paul, Stephen Shifley, and Robert Rogers, *The Ecology and Silvaculture of Oaks.* New York: CABI Publishing, 2002.

Jones, K. and Zell, M. "'The divels speciall instruments': women and witchcraft before the 'great witch-hunt'." *Social History* Vol. 30, N0.1 2005, 45-63.

Jorden, Edward. *A Disease called the suffocation of the mother*, London: John Windes, 1603.

Karlen, Arno. *The Biography of a Germ,* New York: Anchor, 2001.

 DISGUISED AS THE DEVIL

Karlsen, Carol. *The Devil in the Shape of a Woman.* New York: Vintage, 1987.
King James I. (1629). *The Charter of the Massachusetts Bay Company in Essential Documents in American History* at Academic Search Premier http://ebscohost.com.
King William and Queen Mary, "The Massachusetts Bay Charter of 1692" in *Essential Documents in American History* at Academic Search Premier http://ebscohost.com.
Kitron, Uriel. "Predicting the risk of Lyme disease: Habitat suitability." *Emerging Infectious Diseases* 8 No.3, Atlanta, Georgia: Center for Disease Control and Prevention, 2002.
Krech, Shepard. *The Ecological Indian, Myth and History.* New York: W.W. Norton& Co., 1999.
Kurtenbach, K. "The key roles of selection and migration in the ecology of Lyme borreliosis." *International Journal of Medical Microbiology* 291 No.33 June 2002, 152-154.
Lackey, James Alden, David G. Huckaby, Brian Ormiston. "Peromyscus leucopus." *Mammalian Species* No. 247, 1-10. December 13, 1985. The American Society of Mammalogists.
Lang, Denise. *Coping with Lyme disease.* New York: Holt, 2004.
Lapping, Mark, Ed. *Abridged Edition of Howard S. Russell's, The Long Deep Furrow: Three Centuries of Farming in New England.* Hanover, N.H.: University Press of New England, 1982.
Leavitt, Judith and Ronald Numbers, eds. *Sickness and Health in America.* Madison: The University of Wisconsin Press, 1997
Lee, Christopher. *1603.* New York: St. Martin's Press, 2003.
Leigner, K.B. *Lyme Disease Presentation* at the 12th Annual International Conference on Lyme disease and other Spirochetal and Tick Borne Disorders, New York, April 9, 1999.
Leigner, K.B. and C.R. Jones, "Fatal progressive encephalitis." *Abstract* presented at International Conference on Lyme disease and other Tick Borne Diseases at Munich, Germany, June 1999.
Lemay, J.A. Leo. *New England's Annoyances: America's First Folk Song.* Newark, Delaware: Associated University Presses, 1985.
Lesser, R., E. Kornmehl, A. Pachner, J. Kattah, T. Hedges III, N. Newman, P. Ecker and M. Glassman. "Neuro-Opthalmologic manifestations of Lyme disease." *Opthalmology* 97, No. 6, 1990, 699-706.
Lewis, B.A. "Prehistoric juvenile rheumatoid arthritis in a pre-contact Louisiana native population reconsidered." *American Journal Physical Anthropology* 104 June 1998, 229-248.
Lindholt, Paul, Ed., *John Josselyn, Colonial Traveler: A Critical Edition of Two Voyages to New England.* Hanover and London, 1988.
Loftus, John. *The Belarus Secret: The Nazi connection in America.* New York: Knopf, 1982.
Logigian, E., R. Kaplan and A. Steere. "Chronic neurologic manifestations of Lyme Disease." *New England Journal of Medicine*, November 1990, 1438-1444.

# SELECTED BIBLIOGRAPHY

LoGiudice, Kathleen, R. Ostfeld, K. Schmidt, and F. Keesing, "The Ecology of Infectious disease: Effects of Host Diversity and Community Composition on Lyme disease Risk." *PNAS Ecology* 100 No.2, January 21, 2003.

Lyme Disease Association's Position Paper *Conflicts of Interest in Lyme disease:* Laboratory *Testing, Vaccination, and Treatment Guidelines* (April 2001) available at www.lymenet.org.

National Institute of Allergy and Infectious Disease. "Broad- based Vaccination of Wild Mice Could Help Reduce Lyme Disease Risk in Humans." *News Release.* December 4, 2004 at www.niaid.nih.gov.

MacDonald, A. J. Benach and W. Burgdorfer. "Stillbirth following maternal Lyme Disease." *New York State Journal of Medicine*, 1987, 615-616.

MacFarlane, Alan. *Witchcraft in Tudor and Stuart England.* New York, 1970.

Magnarelli, G., K. Stafford III and V. Bladen. "*Borrelia burgdorferi* in *Ixodes dammini* feeding on birds in Lyme, Connecticut." *Canadian Journal of Zoology* 70, 1992, 2322-2325.

Magnarelli, L., John Anderson, W. Adrian Chapell. "Antibodies to Spirochetes in White-Tailed Deer and Prevalence of Infected Ticks from FOCI of Lyme disease in Connecticut." *Journal*" Journal *of Wildlife Diseases* 20 (1) 1984.

Malawista, S.E., A.C. Steere, and J.A. Hardin. "Lyme disease: A unique human model for an infectious etiology of rheumatic disease." *Yale Journal for Biological Medicine* 57, 1984, 473-477.

Markham, Gervase. *Good Cheape Husbandries.* London Reprint of 1614 Edition.

Marshall, W.F., S.R. Telford III, et al. "Detection of Borrelia bugdorferi DNA in Museum specimens of Permyscus leucopus." *Journal of Infectious Diseases.* Vol. 70, 1994, 1027-1032.

*Massachusetts Bay Provision List-* Proportion of provisions needful for such as intend to plant themselves in New England for one whole year. From www.plimouthplantation.org.

Massachusetts Department of Public Health, *Lyme Disease Update for Health Care Providers.* 2002.

Matuschka, F., A. Ohlenbusch, H. Eiffert, D. Richter and A. Spielman. "Antiquity of the Lyme disease spirochete in Europe." *The Lancet* Nov.18, 1995, 1367.

Maugh, Thomas H., "Scientists Use DNA in Search for Answers to Sixth Century Plague." *Los Angeles Times.* May 6, 2002.

McCollum, E. A history of nutrition: the sequences of ideas in nutrition investigations. Houghton Mifflin, Boston, 1957.

McCoy, J. J. "Lyme diagnoses can miss 'bull's-eye': Author Amy Tan finds herself among Unconventional Cases." *Maine Sunday Telegram,* August 17, 2003.

Mennell, Stephen. *All Manner of Food-Eating and Taste in England and France from the Middle Ages to the Present.* Urbana and Chicago: University of Illinois Press, 1996.

Meyer, J.J. and I.L. Bribin, Jr., *Wild Pigs of the United States-their history, morphology and current status.* Athens, Georgia: University Press, 1991.

Michael, Ronald L. Ed. *Historical Archeology,* 32, No.3, 1998.

Millus, S. "Migration may reawaken Lyme disease." *Science News.* February 19, 2000.

Motzkin, Glenn and David Foster. "Grasslands, heathlands, and shrublands in coastal New England: historical interpretations and approaches to conservation." *Journal of Biogeography,* 29, 2000.

Motzkin, G., D.R. Foster, Debra Bernardos and James Cardoza "Wildlife dynamics in a changing landscape." *Journal of Biogeography.* 2002.

Murray, Polly. *The Widening Circle.* New York: St. Martin's Press, 1996.

National Institute of Allergy and Infectious Disease, "Broad- based Vaccination of Wild Mice Could Help Reduce Lyme Disease Risk in Humans." *News Release*, December 4, 2004.

National Park Service Archives, *Revolutionary War Journal of John Ford*, unpublished manuscript in the National Park Service collection, Morristown, N.J.

National Science Foundation. "Secrets of the Woods: Acorns, Biodiversity, and Lyme disease." *News Release.* February 25, 1998 at www.sciencedaily.com.

n.a., "The effects of forest fragmentation on parasiticism of colul monkeys," *The Journal of the Society of American Forestry.* September 2000.

n.a. *Historical Sketches of Andover.* Boston, 1880.

Nicholson, M.C. and T.N. Mather, "Methods for Evaluating Lyme disease risks using GIS and geospatial analysis." *Journal Medical Entomology* 33 No. 5, September 1996.

Norris, Herbert. *Tudor Costume and Fashion.* New York: Dover Publishers, 1997.

Norton, Mary Beth. *In the Devils Snare.* New York: Vintage, 2003.

Oliver, Sandra L. *Saltwater Foodways New Englanders and their food, at sea and ashore, in the nineteenth century.* Mystic, Connecticut: Mystic Seaport Museum, 1995.

Olsen, B., D. Duffy, T.G. Jaenson, A. Gylfe, J. Bonnedahl and S. Bergstrom. "Transhemispheric exchange of Lyme disease spirochetes by seabirds." *Journal of Clinical Microbiology.* Vol.33, No.12, 1995, 3270-3274.

Ostermann, Peter. *Commentarius Juridicus.* Cologne, 1629.

Ostfeld, Richard and Felicia Keesing, "Biodiversity and Disease Risk: the Case of Lyme Disease." *Conservation Biology* 14 No. 3, June 2000.

Ostfeld, Richard, Brian F. Allan, Felicia Keesing, "The Effect of Forest Fragmentation on Lyme Disease Risk." *Conservation Biology* 17, No.1, February 2003.

Ostfeld, Richard. "The Ecology of Lyme disease Risk." *American Scientist.* July-August, 1997.

Parker, J. and K. White. " Lyme borreliosis in cattle and horses: A review of the literature." *Cornell Vet* 82, No.3, 1992, 253-274.

Peel, Dgar Peel and Pat Southern, *The Trials of the Lancashire Witches: A Study of Seventeenth Century Witchcraft.* New York: Taplinger, 1969.

Perley, Sidney. *The History of Salem, Massachusetts.* Salem, 1924.

Pfister, H.-W. , K.M. Einhaupl, V. Preac-Mursic, B. Wilske and G. Schierz. "The spirochital etiology of lymphocytic meningoradiculitis of Bannwarth (Bannwarth's syndrome)."*Journal of Neurology* 231, 1984, 141-144.

Philips, James Duncan. *Salem in the Seventeenth Century* Boston and New York: Houghton Miflin, 1933.

Poland, Gregory A. "Prevention of Lyme disease: A review of the Evidence." *Mayo Clinic Proc.*76, 2001.

# SELECTED BIBLIOGRAPHY

Popvic, N., B. Djuricic, and M. Valcic "The Importance of Lyme borreliosis in Veterinarian Medicine." (Translated from Serbian *Glas Srp Akad Nauka* 43 November 1993.

Pulling, Hazel Adele "CALIFORNIA'S RANGE-CATTLE INDUSTRY: Decimation of the Herds, 1870-1912." *The Journal of San Diego History* January 1965, Volume 11, Number 1.

Qui, W.G., D. Dykhuizen, M. Acosta, and B. Luft, "Geographic Uniformity of the Lyme Disease Spirochete and Its Shared History with Tick Vector in the Northeastern United States." *Genetics* 160, 2002.

Qiu, W.G. *Comparative population genetics of Lyme disease spirochete (Borrelia burgdorferi) and its tick vector (Ixodes scapularis) in North America,* unpublished Ph.D. dissertation, The State University of New York at Stony Brook, 1999.

Rajakumar, Kumarave "Infantile Scurvy: A Historical Perspective." *Pediatrics* 108 No.4, October 2001.

Reik, L. Lyme disease and the nervous system. New York: Thieme (as cited in Fallon, B. and J. Nields) "Lyme disease: A neuropsychiatric illness." *American Journal of Psychiatry.* 1994, 1571-1582.

Richards, Michael P., Rick Sculpting and Robert E. M. Hedges, "Sharp Shift in diet at onset of Neolithic." *Nature* 425, September 25, 2003.

Roach, Marilynne K. *The Salem Witch Trials: A Day-to-Day Chronicle of a Community Under Siege.* New York: Cooper Square Press, 2002.

Marion Roach, *The Roots of Desire.* New York: Bloomsbury USA, 2005.

Rose-Troup, Frances. *Roger Conant and the Early Settlements on the North Shore of Massachusetts.* n.p.: Roger Conant Family Association, 1926.

Rothschild, B.M., K.R. Turner and M.A. Deluca. "Symmetrical erosive peripheral polyarthritis in the late Archaic Period of Alabama." *Science* 241, 1988, 1498-1501.

St. George, R. *Conversing by Signs: Poetics of Implication in Colonial New England Culture* Chapel Hill, N.C.: The University of North Carolina Press, 1998.

Sawyer, Thurman and George Bundren, "Witchcraft, Religious Fanaticism and Schizophrenia—Salem Revisited." *The Early American Review* 3 No. 2, Fall, 2000.

Schmidt, B. E. Aberer, C. Stockenhuber, H. Klade, F. Brier and A. Luger. "Detection of Borrelia burgdorferi DNA by polymerace chain reaction in the urine and breast milk of patients with Lyme borreliosis." *Diagn Microbil Infectious Disease* Vol. 21, 1995, 121-128.

Scott, J.D. "Birds Dispurse *Ixodid* and *Borrelia burgdorferi* infected ticks in Canada." Journal Med. Entomology 38 No.4 July 2001.

Scot, Reginald. *The Discovery of Witches.* Mineola, N.Y.: Dover. Publishing, 1989.

Scott, W.R. *Organizations: Rational, Natural and Open Systems.* Upper Saddle River, N.J.: Prentice Hall, 2003.

Sherman, Paul and Jennifer Billing, "Darwinian Gastronomy: Why we use spices." *Bio Science* Vol 49 No 6, 456.

Shurtleff, Nathaniel. *Records of the Governor and Company of the Massachusetts Bay in New England* Volumes 1, 2, and 3. New York: AMS Press, 1968.

Silverman, Kenneth. *The Life and Times of Cotton Mather.* New York: Columbia University Press, 1971.

Singer, F.J., D.K. Otto, A.R. Tipton, and C.P. Hable, "Home range movements and habitat use of Wild Boar." *Journal of Wildlife Management 5,* 1981.

Sinistrari, Ludovico Maria. (English translation by Montague Summers) *Demoniality*. (London, 1927)

Skaar, Eric. "Did Bloodletting have benefits?" *Science*. September 2004.

Smith, A. G. R. *Science and Society in the Sixteenth and Seventeenth Century*. Norwich, U.K.: Jarrold & Sons, 1972.

Smith, T.B., R.K. Wayne, D.J. Girman, and M.W. Bruford, "A role for ecotones in generating rainforest biodiversity." Science 276, 1997.

Steere, A.C., et. al." Fatal pancarditis in a patient with coexistent Lyme disease and babesiosis. Demonstration of spirochetes in the myocardium." *Annals of Internal Medicine* 103 No.3, September 1985, 374-376.

Steere, A.C., E. Taylor, M.L. Wilson, J.F. Levine, A. Spielman, "Longitudinal assessment of the clinical and epidemiological features of Lyme disease in a defined population." *Journal Infect Disease* 154, 1986.

Steere, A.C., E.J. Dekonenko, V.P. Berardi, L.N. Kravchuk. "Lyme borreliosis in the Soviet Union." *The Journal of Infectious Disease* 158 No.4.October, 1988.

Steere, A.C. and E. Dwyer, and R. Winchester, "Association of Chronic Lyme arthritis with HLA-DR4 and HLA-DR2 alleles." *The New England Journal of Medicine* 323, July 26, 1990.

Stewart, Angela. 28. Angela Stewart, "Study finds changes in Lyme bacteria." *The Star Ledger*. September 28, 2004.

Stratton, Eugene. *Plymouth Colony: its History and People 1620-1691*. Ancestry Publishing: Salt Lake City, Utah: Ancestry Publishing, 1986.

Subak, Susan. "Effects of Climate Variability in Lyme disease Incidence in the Northeastern United States." *American Journal of Epidemiology* Vol.157, Number 6, 2003, 531-538.

Summers, Montague, Ed. *The Malleus Maleficarum of Heinrich Kramer and James Sprenger*. (Mineola, N.Y.: Dover, 1971) available online at http://malleusmaleficarum.org

Summerton, Nicholas. "Lyme disease in the 18th Century" British *Medical Journal* 311, December 2, 1995, 1478.

Tabor, Clarence Wilber. *Tabors Cyclopedic Medical Dictionary*. F.A. Davis, Philadelphia, 1952.

Terborgh, J. "The role of ecotones in the distribution of Andean birds." *Ecology* 66, 1985.

Thomas, Clayton, ed. *Tabors Cyclopedic Medical Dictionary*. F.A. Davis, Philadelphia 1997.

Thomas Tusser, *Five Hundred Points of Good Husbandrie*. London, 1577 Reprint 1878.

United States Code. *The Federal Clean Air Act*. 1970. Title 42-Chapter 85.

United States Department of Agriculture, *Vitamin C content in Foods Nutrient Database*, Release 12, 1998

_____, Soil Conservation Service. *Soil Survey of Barnstable County, Essex County: the southern part* and *Plymouth County in Massachusetts*. September 1986.

United States Department of Health and Human Services, *Healthy People 2010*. Washington, D.C.: Government Printing Offices, 2000.

University of Virginia. *Salem Witchcraft GIS* project information can be found at http://lewis.lib.virginia.edu

Upham, W.P. *Salem Witchcraft* Vol. I and II. Boston, 1867.

# SELECTED BIBLIOGRAPHY

Vanderhoof-Forschner, Karen. *Everything You Need to Know about Lyme disease and other tick-borne disorders.* New York: John Wiley & Sons, Inc., 1997.

van Hoof, Thomas, Frans Bunnik, et al. "Forest re-growth on medieval farmland after the Black Death pandemic-Implications for atmospheric $CO_2$ levels." *Palaeogeography, Palaeoclimatology, Palaeoecology,* 237 (2006), 396-411

Watt, J., E. J. Freeman, and W. F. Bynum. *Starving Sailors: The Influence of Nutrition upon Naval and Maritime History.* London: National Maritime Museum, 1981

Weber, K. and W. Burgdorfer. "The cradle of Lyme borreliosis." *Journal of Spirochete and Tick –borne disorders.* Vol. 1, No. 2, 1994, 35-36.

Weber, M. (1992 reprint). *The Protestant Ethic and the Spirit of Capitalism.* Talcott Parsons: London, 1992, p 61.

Wehrwein, Peter. "Nantucket Fever: Entomologist Andy Spielman's Search for the Creeping Carrier of Babesiosis, Ehrlichiosis and Lyme disease." *The Harvard Public Health Review* at www.hsph.harvard.edu

Weiner, Michael and Janet Weiner. *Herbs that Heal.* Quantum Books, Mill Valley, California, 1994.

Weissman, K., L. Jagminas, M.J. Shapiro. "Frightening dreams and spells: a ventricular asystole from Lyme disease." *European Journal Emergency Medicine* 6, December 1999, 397-401.

Wessels, Tom. *Reading the Forested Landscape: A Natural History of New England.* Woodstock, Vermont: Countryman Press, 2003.

Willison, George. *Saints and Strangers.* Orleans, Massachusetts: Parnassus Imprints, Inc., 1945.

Wilson, M.H. "Reduced abundance of adult Ixodes dammini (Acari: Ixodidae) following destruction of vegetation." *Journal of Economic Entomology*, Vol. 79, 1986, 693-696.

Winslow, Edward. *Mourt's Relation.* Bedford, Massachusetts: Applewood Books, 1963.

Winthrop, John. A Model of Christian Charity Sermon, 1630, *in Essential Documents in American History* at Academic search Premier http://ebscohost.com.

Winthrop, John. *The Journal of John Winthrop 1630-1649.* Cambridge, Massachusetts: Belknap Press, 1996.

Winthrop, John, Jr. *Letters of John Winthrop, Jr.* Massachusetts Historical Society Collections. unpublished.

Wood, William. *New England's Prospect.* Amherst: University of Massachusetts Press, 1977.

World Health Organization *SCURVY and its prevention and control in major emergencies* A review was prepared by Zita Weise Prinzo, Technical Officer, *WHO,* 1999.

Zennie, Thomas and C. Dwayne Ogzewalla. "Ascorbic acid and Vitamin Content of Edible Wild Plants." *Journal of Economic Botany,* 1977.

DISGUISED AS THE DEVIL

# INDEX

## A

*A Dialogue of Witches*, 57, 265
*A Fever in Salem*, 69, 75
*A Model of Christian Charity*, 189, 202
acorn poisoning, 140, 149, 198, 225
acorns, 71, 80, 96, 123, 144, 270
acrodermatitis chronica atrophicans, 15, 56
Africa, 19, 21, 29, 48, 73, 158
African Swine Fever, 48
Agobard, the Archbishop of Leon, 153
Alabama, 49, 73, 271
alacrity, 20
Albigensian Crusades, 155
Alps, 92
American Academy of Neurology, 211, 213
Andover, 140, 149, 167, 176, 178, 179, 180, 185, 205, 252, 254, 255
anesthetic, 23, 24, 32, 205
antibiotics, 4, 5, 14, 15, 27, 43, 53, 117, 147, 207, 214
antimicrobial activity, 41
anti-Semitic, 155
Arctic Circle, 19
armor, 83, 118, 133
Aronowitz, Robert, 14, 28
arthritis, 15, 27, 41, 43, 50, 53, 64, 73, 121, 268, 272
Asia, 49, 79
Asymptomatic, 21
Atkins' diet, 111
Atlantic Ocean, 188
Attorney General of Connecticut, 211
autumn, 80, 81, 104, 198
Azore Islands, 93

## B

babesiosis, 20, 128, 272, 273
Baltimore, 84
Bannworth's syndrome, 15
Barbados Islands, 50
bartonella, 20, 31, 61

Battle of the Boyne, 197, 202

Bayts, Margaret (accused witch), 160
beans, 129, 229
bedbugs, 62
beef, 106, 109, 111, 130, 131, 132, 198, 224
beer, 93, 107, 110, 111, 204
Behringer, Wolfgang, 73, 92, 95, 96, 158, 171, 172, 236
Bell's Palsy, 16, 43, 45, 54, 63, 65
Bergesen, Albert, 188, 202
Bering Strait, 49
bewitched, 64, 70, 102, 159, 162, 165, 168, 169
Big Dig, 51, 109, 145
biodiversity, 35, 71, 96, 270
bio-weapon research, 217
birds, 18, 29, 31, 32, 34, 36, 37, 39, 47, 48, 50, 72-74, 77, 81, 138, 139, 144, 147, 205, 225, 267, 269, 272
bite mark, 60, 205
black witches, 157
blackbird, 50
blood meal, 19- 22, 30, 32, 35, 47-49, 80, 81, 119, 160
Bodin, Jean, 157, 171
Borrelia burgdorferi, 4, 8, 15, 18, 20-22, 29, 31-33, 35, 37, 41-48, 50, 53, 56, 68, 71-74, 80, 117, 121-123, 209, 216, 226, 264, 269, 271
Boston, 11, 18, 28, 37, 50, 62, 69, 70, 74, 76, 83, 94, 109, 112, 113, 122, 125, 128, 132-135, 143-145, 149, 162, 164, 172, 176, 177, 179, 202, 204, 206, 217-219, 248, 250
Bradford, William, 71, 83, 107-112, 115, 116, 118-120, 122, 124, 135-137, 144, 149, 257, 264
Brewster, William, 122
broadsides(to advertise witch burning), 155
brood parasitism, 37
bubonic plague, 152
bull's eye rash, 14, 15, 25, 56, 205
Dr. Burgdorfer, Wily, 4, 44
burning, 39, 61, 75, 112, 125, 132, 146, 147, 164, 182, 198, 200, 215, 254
burning executions, 164
Bush, George W., 6, 27, 55

## C

Calef, Robert, 11, 28, 63
calendar, 25, 104, 140
California, 42, 74, 127, 128, 138, 139, 141, 143, 144, 149, 264, 273

Canada, 29, 49, 63, 74, 271
canine, 46, 83, 84, 95
canoes, 176
Cape Ann, 176
Cape Cod, 37, 46, 78, 83, 92-95, 107, 109, 111, 113, 118, 119, 122, 123, 127, 132, 232
Caporeal, Linnda, 68
Carr, James, 70, 76
Carlson, Laurie Winn, 69, 76, 264
caste young. *See* spontaneous abortion
Cathar, 154, 155, 171
cathelicidin antimicrobial peptide gene, 41
Catholic Church, 153, 187, 188, 193, 195, 196
cats, 13, 22, 29, 46, 164, 208, 215
cattle, 48, 56, 109, 111, 112, 126-129, 131, 135, 137, 143, 152, 168, 177, 180, 198, 225, 270
CDC (Center for Disease control), 14, 27, 74, 92, 96, 208, 214
cell wall deficient, 20, 24, 44
Chandler, Phoebe, 66, 178, 254
charter, 189, 192, 193, 195, 196, 197
Cherokee folklore, 53
children, 69, 119, 168, 250
chimney, 166, 173, 194
chipmunks, 7
chronic Lyme disease, 7, 8, 208-213
Civil War, 51, 52, 74, 267
clams, 113
Clark, William, 52
class action lawsuit, 6
climate, 18, 19, 35, 57, 71, 77, 92, 93, 128, 152, 195, 267
cold temperatures, 159
Colorado Tick Fever, 52
Columbia University, 271
Columbus, Christopher, 50, 73, 270
*Commentarius Juridicus*, 60, 75, 270
commercial laboratory testing, 209
common land, 177
*Compendium Maleficarum*, 65, 75, 93, 103, 159, 168
Conant, Roger, 176, 185
conflicts of interest, 3, 6, 211
congenital transfer, 46, 119, 207
Congress, 5, 202, 217
Connecticut, 12, 14, 15, 22, 28, 37, 70, 112, 115, 121, 126-128, 135, 161, 167, 175, 176, 180, 188, 206, 269, 270
Corey, Giles and Martha, 167, 172, 220, 255

corn, 108, 110, 127, 129, 130, 142, 147, 158, 225
Corwin, George, 131
Cotta, John,163, 168, 172, 173, 264
Cotton, John, 189
covenant, 187, 189, 191, 193
cows, 22, 46, 129, 225, 251
Creek folklore, 53
Cross Street privy, 133
Crusades, 106, 155, 171
culture, 10, 75, 103, 172, 202
curse, 17, 163
cyst, 20, 24, 44
Czech Republic, 92

# D

*Daimonomageia*, 164, 172
Daneau, Lambert, 57, 265
Danvers, 1, 176, 179
day of prayer and fasting, 94
de Montaigne, Michel, 157, 162, 171, 172
deaths, 5, 6, 107, 111
deciduous forests, 20, 77, 82
deer, 15, 17, 18, 19, 21, 31, 32, 37, 39, 40, 47, 53, 61, 71, 73, 77, 79-85, 93, 107, 109, 110, 112, 117- 119, 123, 125, 129-133, 135, 138, 139, 143-147, 153, 161, 200, 201, 205, 216, 225,226, 267
DEET, 8, 28, 30, 40
deforestation, 15, 37, 152, 153, 158, 178, 179, 204
dementia, 43
*Democracy as a Detriment to church and state*, 189
Demos, John, 67, 75
dermatological research, dermatologists, 14, 15
Desboroughs, Mercy (Marcy) (accused witch), 167
devil's mark, 57-60, 160, 200
*diablo stimata*, 57
diet, 16, 30, 41, 80, 105, 106-112, 115, 117, 130, 133, 224, 271
*Disquisitionum Magicarum*, 58, 75
dissembling, 169
disseminated Lyme disease, 45
distraction, 169
division between church and state, 207
DNA, 15, 29, 44, 47, 48, 53, 57, 74, 206, 219, 269, 271
Doctrine of Effluvia, 162

# INDEX

dogs, 7, 8, 13, 22, 29, 46, 48, 49, 73, 84, 129, 138, 164, 168, 226
domesticated animals, 2, 16, 22, 46, 109, 133, 145, 158, 207, 216
drag test, 119, 123
drought, 19, 21, 22, 29, 30, 68, 82, 93-95, 158, 159, 171, 192, 193, 197-199, 205, 231, 232
Duke of Savoy, 155
Durkeim, Emile, 187, 265
Dutch North Sea Coast, 93
Dyer, Mary, 76, 102, 242

## E

earmarks (ownership of animals), 127, 138, 149
ECM, 5, 8, 14, 15, 30, 43, 56-61, 63, 110, 161, 183, 202, 227
ecotones, 29, 35, 38, 50, 71, 74, 184, 204
Elbe River, 152
ELISA blood test, 5
Elliot, Daniel, 169
encephalitis, 27, 46, 69, 268
Encephalitis Lethargica, 69
encephalomyelis, 46, 66
endemic areas, 6, 22, 42, 46, 208, 215
Endicott, John, 176, 177, 178, 179
English, (colonists) 16, 17, 37, 48, 50, 69, 105, 106, 107, 108, 111, 112, 117, 126, 128, 129, 130, 133, 137, 140, 143-146, 157, 164-166, 171, 175, 176, 178-180, 187, 188, 191, 193, 194, 197, 198, 206, 208, 209, 231, 232, 265, 272
environmental, 1, 2, 13, 18, 19, 22, 35, 38, 44, 48, 84, 96, 153, 197, 203, 232
ergot, 67, 68, 75, 264
Erikson, Kai T.,195
Europe, 1, 13, 14, 17, 21, 30, 48-50, 52, 55, 56, 58, 61, 64, 75, 79, 80, 85, 88, 95, 106-108, 122, 139, 152, 156-158, 164, 171, 173, 188, 191, 193, 194, 203, 206, 265, 269
european population levels, 152
evil eye, 97, 162, 168, 172

## F

Dr. Fallon, Brian, 44
farm abandonment, 37, 153
farmers, 42, 131, 140, 198
fatigue, 17, 28, 43, 45, 54, 55, 63, 66, 106, 112, 115, 119, 224, 254
FDA(Food and Drug Administration), 6, 7, 46
ferrets, 206
fertility, 167, 176, 205
fetal deformity, 46
fetuses, 70
fever, 15, 18, 21, 43, 48, 100, 112, 121, 125, 127, 128
fibromyalgia, 55
Fifth Amendment (U.S. Constitution), 202, 209, 220
fish, 41, 100, 107-111, 127, 129, 146
fit(s), 16, 37, 62, 68, 101, 108, 117, 128, 162, 169, 180, 195, 204, 207
Fitch, Asa, 52
Five Mile Pond, 166, 198
Florida, 29, 37, 55, 207
flu-like symptoms, 43
flying through the air, 167
folkloric, 17, 152, 164,166
Dr. Fontaine, Jacques, 59, 75
for profit health care system, 213
forest fragmentation, 37, 71, 91, 270
fossilized tick, 47
founding generation, 187, 191, 192
France, 115, 154, 157, 203, 206
Freud, Sigmund, 18, 28
frontier towns, 131

## G

genome sequence, 33
geography of risk, 39
Georgian Mindset, 194
germ theory of disease, 25
Germany, 27, 88, 92, 154-157, 203, 236, 268
GIS, 83, 84, 91, 95, 270, 273
glacial, 19, 30, 35, 77
global warming, 19, 22, 29
goats, 46, 106, 126, 137, 140, 143, 152, 155
Good, Dorcas (accused witch), 61
Good, Sarah (accused witch), 62, 140, 149, 199,200, 253, 269, 272
*Good Cheape Husbandries,* 140, 149
Goodwin, children, 12, 61-66, 164, 199
Gookin, Daniel, 71, 135, 175, 185, 266
government, 6, 189, 190-193, 195, 196, 207, 210
grain-centric mind, 110

grass, 16, 107, 113, 117, 119, 126, 129, 143, 200, 229
Great Famine of 1315, 152
groundnuts (Jerusalem artichokes), 110
Gulf Coast, 35

## H

habitat, 31, 32, 35, 38-40, 77, 81, 85, 129, 139, 147, 149, 201, 226, 272
halfway covenant, 191
hallucinations, 16, 45, 50, 55, 63, 66, 169
hand of man, 32, 36
Hansen, Chadwick, 68, 76
Hartford, Conn., 127, 161, 243
Harvard University, 75, 95, 124, 190, 219
Henry Tudor, 48
herbs and spices, 17, 229-230
heresy, heretics, 153, 155-157, 188
Hertford, 164
Higginson, Frances, 71, 112, 113, 116, 135, 137, 149, 176, 185, 239
high risk for Lyme disease, 18, 83, 84, 91, 118, 179
*Historia de Gentibus septentrionalibus*, 57, 74
HLA-DR2, 3, 6, 27
HLA-DR4, 3, 6, 27
hog, see also pig, 130, 146
homeland security, 217
horse, 123, 168, 176, 199
H(F)ortado, Mary, 17, 60, 66, 75, 199
house pets, 208
Hubbard, Elizabeth, 169
human bite (mark), 16
hung at sea, 166, 173
Hungary, 93
hunting, 37, 42, 47, 85, 107, 109, 110, 130, 132, 133, 135, 138, 139, 146, 147, 156, 171, 193, 200, 205
Hutchinson, Ann, 102, 242
Hutchinson, Thomas, 11
hysteria, 67, 69, 157, 169, 173, 196, 267

## I

IDSA (Infectious Disease Society of America), 7, 27, 208, 210, 211, 212, 214, 217
IDSA guidelines, 8, 211, 214
ILADS, 7, 27, 209, 210, 212
immanent, 188, 190, 194
immune response, 44, 121

immunosuppression, 24, 44-45
impaired concentration, 56
*In the Devil's Snare*, 67, 69, 172, 219
indoor lifestyle, 25
Industrial Revolution, 22, 153
infect-obesity, 200
infertility, 99, 158
insecticides, 17, 25
Ipswich River, 21, 94, 177, 178, 180, 181
Ipswich Road, 177, 178
Ireland, 28, 157, 164, 197, 236
Italy, 46
ivy, 17, 164
*Ixodes* family of tick, 10, 15, 19-22, 25, 29, 31, 32, 34, 35, 37, 39, 41, 45, 47-49, 53, 56, 61, 62, 71, 77, 78, 80, 83, 95, 117, 119, 123, 124, 143, 148, 150, 171, 203, 204, 224, 264, 266, 269, 271, 273
*Ixodes dammini*, 34

## J

Jackson, Andrew, 52, 74, 95
Jamestown, 50, 74, 108, 109, 224, 239
Jefferson, Thomas, 195, 202
joint pain, 15, 43, 46, 51, 101, 112, 119, 224, 239
judges, 57, 157
Jura, Island (Scotland), 49, 73
Justinian's plague, 206

## K

Kalm, Peter, 51, 71, 74
Kant, I., 193
Karlen, Arno, 56, 75
Katherine Nanny privy, 109
King James I, 167, 188, 202
King James II, 197
King William, 193, 197, 202
kiss of the cat, 155
Knapp, Elizabeth, 12, 28, 64, 66, 199
Knopp, Nicholas, 114, 117
Kramer, H., 156, 171, 272

## L

land grants, 177, 178
land scurvy, 17, 18
landscape, 16, 32, 51, 71, 79, 81, 82, 89, 91, 94, 111, 117-119, 125, 128, 129, 135, 143, 152, 170, 175, 176, 179, 200, 204, 270

# INDEX

lard, 40
Larner, Christina, 57, 74, 75
larvae, 31
Laurasia, 47
Lawson, Deodat, 61
leaf litter, 31, 32, 39, 79, 80, 128, 143, 144
leeks, 113
lemon, 107, 112, 116, 117, 229
lemon grass, 107
Lewis and Clark Expedition, Corps of Discovery, 52
Lewis, Mercy, 11, 207
Little Ice Age, 57, 92, 93, 96, 104, 158, 159, 171
livestock, 131
lizards, 35
lobster, 110
low risk for Lyme disease, 83, 84
Lyme Activists, 4
Lyme literate physician, 216
lymphocytoma, 15, 56, 60

memory loss, 46, 55
meningitis, 43, 66, 101
menstruation, 44
mental illness, 28, 43, 67, 169, 215
Middle East, 152
Midwest, 35, 50, 207
militia, 179, 190, 193, 197, 198
milk, 271
mindset, 62, 92, 106, 108, 175, 193
miscarriage, 70, 183
modern medicine, 216
moisture, 21, 22, 82, 84, 93, 94, 153
*More Wonders of the Invisible World*, 11
mosquitoes, 51, 69
mountain lion, 129
mouse, 42, 47, 52, 80, 81, 82, 84, 129, 143, 153, 205, 215, 216
MRI scans, 56
Murray, Margaret, 60
Murray, Polly, 18, 29, 54, 70, 74, 209
musket, 130, 132, 133

## M

*Macbeth*, 167
MacFarlane, Alan, 203, 219
madness, 29, 157, 173
Maine, 17, 25, 48, 61, 74, 76, 108, 116, 123, 124, 165, 192, 193, 205, 269
male apparel, 118
*Malleus* (Hammer of Witches), 156, 157, 171, 173
market, 6, 129, 131, 132, 216
Markham, Gervase, 139, 149
Martin, Susannah, 29, 73, 168, 220
Maryland, 53, 82, 84, 94, 231, 232, 266
Massachusetts, 11-13, 17, 18, 21, 35, 38, 39, 47, 51, 71, 72, 75, 78, 79, 82, 90, 91, 92, 95, 96, 105, 106, 109, 111-117, 119, 123-133, 135, 137-140, 142-146, 149, 158, 159, 175-180, 185, 187-197, 202, 206, 207, 209, 220, 258, 263, 265-273
mast (acorns), 39, 79, 80, 81, 140, 142, 147, 198, 199
masting, 22, 80, 84, 93, 153
*Mayflower*, 107, 118, 119, 121, 137, 148
Mayo Clinic, 271
McBride, William, 52
MC1R gene, 164
medical insurance, 4, 212, 214
medieval settlements, 152
Medieval warm period, 152
meeting house, 167, 192

## N

Native American, 15, 17, 50, 53, 81, 112, 122, 123, 129, 132, 135, 146, 175, 177, 179, 182, 192, 203, 205, 231, 232
Native Americans, 182
Naumkeag, 176
NEJM (New England Journal of Medicine), 211
neurasthenia, 10, 17, 28
neurological symptoms, 2, 11, 16, 21, 25, 64-67
New England, 3, 11, 12, 16, 25, 27, 28, 31, 32, 36, 37, 46, 49, 71, 73, 74, 75, 76, 80, 84, 85, 92, 94, 95, 107, 115, 116, 117, 123, 128-130, 133, 135, 137, 143, 144, 147, 149, 150, 156, 158, 161, 163, 165, 168, 169, 171, 173, 175, 182, 185, 189, 190, 191, 194, 195, 197, 199, 202, 205, 206, 208, 209, 212, 216, 220, 232, 258, 263-273
New Hampshire, 83, 175, 207, 231
New Haven, Conn., 76, 209
New Jersey, 47, 55, 67, 72
New World, 12, 50, 73, 105, 109, 122, 130, 133, 137, 152, 159, 165, 166, 191, 264
New York, 13, 22, 27-29, 55, 71-76, 83, 95, 103, 104, 115, 116, 124, 135, 136, 149, 171-173, 185, 202, 219, 220, 231, 252, 264, 271
NIH, 210, 215

Norman Conquest of 1066, 203
North America, 1, 2, 17, 19, 28, 31, 33, 34, 35, 38, 47, 48, 50- 52, 71, 72, 79, 80, 86, 87, 89, 106, 158, 171, 173, 271
Northern hemisphere, 159
northern lineage(ticks), 35
Norton, Mary Beth, 67, 69, 220
nymphal tick, 29, 81, 82, 89

## O

oak, 30, 79, 83, 140
*Odocoileus virginiansis,* see white-tailed deer
Oligocene, 47
*On the Demon-Mania of Sorcerers,* 157, 171
onions, 113
opossums, 36
Orchard Farm, 177, 178
*Ordering Their Private World,* 190, 202
OspA, 7, 27
Ostermann, Peter, 60, 75, 270
oyster, 113, 229

## P

pagan, 75, 152, 155
Paisley, Brad, 58, 75
pandemic, 267
Parris, Betty, 12, 28
passenger pigeons, 123, 143, 144
pasteurization, 46
penicillin, 14
Pennsylvania, 49, 75, 82, 171, 173, 231, 232, 267
*peromyscus leucopus,* 47, 74, 268
Phips, William, 193
photophobia, 97, 162
physician(s), 7, 18, 28, 97, 163, 168, 169, 216
pig skeletal remains, 51
pigs, see also hog, 22, 137, 139, 140, 145, 149, 264, 269
Pilgrims 104, 107-109, 112, 117, 119, 120, 125, 130, 132
Pinnacles National Monument, 138
pin(s), 16, 61- 64, 165, 166, 204, 224, 227
Pleiocene Epoch, 47
Pleistocene, 35, 78
Plum Island, 13, 28, 264
Plymouth, 17, 37, 71, 78, 91, 94, 105, 107, 108, 109, 110, 111, 112, 115, 116, 121, 122, 124-126, 128, 129, 131, 133, 135-137, 143, 144, 149, 175, 176, 185, 188, 191, 206, 207, 224, 257, 264, 265, 267, 272
pollen analysis, 82, 83, 91, 95, 142, 177
pollen core sampling, 142, 177
Pope Innocent III, 155
population contraction, 35
pork, 106, 130, 145
Post Lyme Syndrome, 55
potash, 176
predators, 85, 129
Proctor, family, 167
Protestant Reformation, 155, 195
PRZ, 143, 144, 145, 146
psychiatric Ward, 70
psychosis, 66
Public Health, 39, 42, 72, 79, 95, 124, 219, 266, 269, 273
puffins, 48
pumpkins, 129
Puritans, 12, 104, 109, 125, 130, 131, 159, 187, 188, 190, 196, 202, 266
Putnam family, 177-179, 182, 183, 192
Putnam, Ann, 12, 28, 63, 173, 183, 199

## Q

Quebec, 193, 197- 199, 205
Quebec invasion of 1690, 193
Queen Mary, 197, 202
quest (ticks), 32, 80, 81, 212

## R

rainfall, 21, 92, 158, 159, 231, 232
rats, 27, 48, 206
Reagan, Ronald, 189
red haired, 13, 164
reduced genetic variability, 35
reeves, 130, 135, 146
re-mystification campaign, 155
research monies, 10
Revolutionary War, 37, 51, 74, 267, 270
rheumatism, 10, 51-53, 226
rheumatologists, 14
Rhine River, 203
Rhone River, 203
Riegate Witch Bottle, 165
rifle, 133

# INDEX

riparian, 21, 30, 31, 77, 82, 84, 92, 158, 179, 184, 231
risk factors, 266
robins, 50
Rome, 153
rose hips, 107, 113, 126, 229
Rule, Margaret, 61, 63, 65, 199

## S

sabbat, 155
Saint Augustine of Hippo, 153
*Salem Possessed*, 12, 67, 75, 219, 263
Salem Village, 1, 11, 21, 61, 68, 69, 90, 93, 126, 130, 140, 148, 165, 167, 169, 170, 172, 175, 176, 179, 180, 185, 192, 193, 195, 197, 198, 199, 206, 207, 208, 219, 232, 252, 254, 255
Salem Witch Trials, 1, 16, 53, 67, 76, 219, 266, 271
Santa Claus, 167, 173, 194
sassafras, 83, 107, 108, 229
Satan, 76, 153-155
science, 3, 8, 10, 13, 25, 29, 77, 151, 188, 191, 193, 194, 209
scientific revolution, 191
Scotland, 49, 58, 75, 191
Scots-Irish, 49
Scott, Reginald, 151, 163
scurvy, 10, 16, 17, 18, 48, 105- 107, 110, 112, 113, 115, 120, 125, 126, 128, 204
seizures, 10, 11, 16, 45
seropositivity, 22, 42, 94, 208
settlers, 16, 18, 32, 36, 37, 50, 74, 105, 107-112, 114, 117, 119, 122, 125, 127-130, 132, 133, 137, 141, 142, 166, 175-177, 179, 180, 191, 194, 198, 204, 206, 239
Sewall, Samuel, 68, 76
Sexton, Wyatt, 54
Shakespeare, William, 167, 173, 196
shape shifting, 15, 44, 48
Shawshin River, 166, 178, 180
sheep, 106, 131, 137, 143, 152, 220
shellfish, 110, 113
Short, Mercy, 11, 62-67,199
shrews, 7
sickness, 10, 15, 16, 20, 39, 50, 73, 97, 105, 111, 112, 121, 125, 126, 128, 140, 144, 166, 168, 180, 182, 251
skepticism, 14, 157, 194
smallpox, 2, 129, 193

Smith, John, 6, 71, 104, 115, 143, 149, 164, 176, 185, 202, 239
Smith Kline Glaxo, 6
snapchance, 118, 133
social boundary, 195
soil, 5, 21, 77, 78, 91, 95, 126, 139, 177, 179, 183, 185, 272
sorcery, 163, 171
Soviet Union, 10, 27, 43, 72
spectral visions, 169
spirochete, 4, 5, 15, 18, 20, 27, 32, 40, 44, 45, 49, 71, 81, 115, 121, 208, 209, 214, 266, 269, 271
spontaneous abortion, 46, 111, 112, 127, 207
Sprenger, J., 156
squirrels, 36, 138
Standish, Miles, 118, 122
star phylogenies, 35
starvation, 105
starving times, 16, 74, 108, 111
*stigmata diaboli*, 160
stinking suffumigations, 165
strawberries, 18, 113
streptococcus, 121
stress, 20, 22, 24, 29, 34, 44, 74, 122, 192, 205, 232
suburban, 2, 37, 84, 85, 147, 153
Sudbury, Mass., 125, 131, 132
sunken Spanish treasury ship, 193
survivors, 47, 121
Sweden, 19, 49, 73, 77, 95
swine, *see also hog, pig*, 48, 129, 131, 137-143, 146
Switzerland, 92, 154
swollen glands, 45

## T

Tan, Amy, 55, 70, 74, 76, 269
tariff, 131
Tchefuncte Indian adolescent skeleton, 50
temperature, 34, 35, 47, 50, 60, 92, 145, 158
Tennessee, 54, 75
terriers, 206
Texas or Southern fever, 127
*The Biography of a Germ*, 56, 74
The Black Death, 57
the disenchantment of the world, 151, 188, 194
the Enlightenment, 10, 151, 194
*the Impact of the Little Ice Age on Mentalities*, 92, 96

## DISGUISED AS THE DEVIL

the Inquisition, 155
the mother (hysteria), 25, 76, 169, 173, 267
The Sweating Fever, 48
*The Widening Circle,* 29, 55, 74, 270
theocracy, 189, 196
theories, 10, 28, 195
tick life cycle, 23
tick or famine fever, 21
tinnitus, 28, 66
tree rings, 68, 92, 93
trigger, 22, 41, 44
tundra, 267
Tuskegee, 5
Tusser, Thomas, 140, 143, 149, 272

## U

USDA, (U.S. Department of Agriculture), 139, 264, 272
University of California, Irvine, 218
University of North Carolina, 45, 74, 115, 148, 202
urban rats, 7
urine, 271

## V

vaccination, 25, 27, 220
vaccine, 5, 6, 7, 44, 46, 214, 215, 217
venison, 109, 110, 132, 133
vertebrate reservoir hosts, 48
veterinarians, 46, 214, 216
Vikings, 49, 73, 137, 148, 266
Virginia, 16, 50, 74, 91, 95, 108, 111, 115, 129, 149, 185, 219, 263, 265, 273
vitamin C, 17, 18, 106, 107, 111-113, 117, 125, 126
vitamin D, 40-42, 109, 111, 122
volcanic, 158, 159

## W

Waldensians, heretics, 154
Walker, John, 49
War, 8, 51, 74, 127, 130, 177, 192, 193, 205, 231
Warren, Mary, 169

water, 16, 20, 21, 25, 40, 44, 82, 84, 94, 100-111, 114, 127, 160, 167, 173, 176, 177, 198, 226
weather making, 92
weather patterns, 159
Weber, Max, 151, 187, 194
weeds, 39, 123
West Indies, 131, 132
Westchester County, 83, 84
WESTERN BLOT blood test, 5
Weston, Thomas, 111
white footed mice, see *peromyscus leucopus*
white tailed deer, 20, 31, 36, 117, 123, 124, 128, 143, 144, 147, 198, 264
wilderness, 13, 128, 129, 140, 175, 176, 178, 189
Wilkins, Bray, 177, 252
William Wood, 112, 116, 149, 185
Williams, Abigail, 12, 28, 199
will-o-wisp, 168
Winslow, Edward, 39, 50, 71, 107, 108, 109, 111, 112, 115, 119, 122, 124, 137
Winthrop, John, 50, 76, 109, 112, 115, 116, 125, 126, 127, 129, 131, 135, 142, 143, 149, 180, 185, 189, 195, 273
Wisconsin, 14, 39, 77, 78, 135
witch, 1, 13, 17, 18, 28-30, 57-63, 66, 120, 153, 156-158, 160-167, 171-173, 191, 192, 195, 196, 199, 207, 217, 227, 231, 267
witch-hunting, 156
witch's teat, 13, 30, 60
witchcraft affliction, 10, 62, 156, 164, 165, 204
*Witchcraft at Salem,* 68, 76
Witchcraft Hysteria, 207
*Witchcraft in Tudor and Stuart England,* 203, 219
witchcraft persecution, 89
witchcraft prosecutions, 49
Witchcraft Reparations of 1704 (Salem), 194
witchcraft trials, 12, 158, 166, 190
wolf, 129, 130, 139, 142, 216, 226, 267
women's clothing, 119
wool, 118, 131, 178
working outdoors, 42

# INDEX

## Y

Yale University, 14, 76, 121
yersinia pestis, 206

## Z

zooinosis model, 48

## DISGUISED AS THE DEVIL

# ACKNOWLEDGEMENTS

Any historical work is in many ways a collaborative process. The past contributes information in a written form while the present is available for interactive communication. My family has always supported my interests and kindly tended to my ailments as best they could. My mother has always offered support, even as her own health and memory fades. My beloved spouse has been supportive from the start of this project and my children have offered their expertise, especially when it comes to graphics and computer technology. My sister has always served as my best sounding board. The many people that I have come into contact with in the Lyme afflicted community have offered me inspiration for a number of years now. The Lyme literate physicians who work exhaustively, often under the adverse political conditions created by our health care system, to improve the lives of their patients should be treated like true heroes. The academic support that I received from USM while studying in the American and New England Studies Program and during my doctoral training is well appreciated. I would particularly like to thank Kent Ryden and Ardis Cameron, who served as my Master's thesis advisor and reader. I would also like to thank Mark Lapping for exposing me to new and different ways of seeing the world, and to Barbara Fraumeni for making numbers make some sense to me for the first time in my life. The support staff at the Glickman Library, especially the Reference and Special Collections Librarians, have been extraordinarily helpful. The research and work that was done in the past on the witchcraft records of Salem has been exemplary. I feel like I have walked in the shadow of those before me who have been attracted to this subject for well over three hundred years. Moreover, we all owe a debt to the WPA workers who toiled to carefully translate hand written, often difficult to read, documents. The University of Virginia's Salem Witch Trials Documentary Archives and Transcription Project that has made these original documents accessible electronically to a wide audience has created an invaluable resource that I have used extensively over the past many years. Wolfgang Behringer's extensive work *Witches and Witch-Hunts* has also been inspirational.

South Portland, Maine<br>
March 2008

DISGUISED AS THE DEVIL

www.ingramcontent.com/pod-product-compliance
Lightning Source LLC
Chambersburg PA
CBHW032035150426
43194CB00006B/293